Progress in Precision Agriculture

This book series aims to provide a coherent framework to cover the multidisciplinary subject of Precision Agriculture (PA), including technological, agronomic, economic and sustainability issues of this subject. The target audience is varied and will be aimed at many groups working within PA including agricultural design engineers, agricultural economists, sensor specialists and agricultural statisticians. All volumes will be peer reviewed by an international advisory board.

Paul Newell Price • Lizzie Sagoo
Roger Robson-Williams
Editors

Precision Agriculture in Field Horticulture

The Potential for Agri-technology to Facilitate Innovation and Promote Value in Outdoor Fruit and Vegetable Production

Springer

Editors
Paul Newell Price
ADAS
Sustainable Agricultural Systems
Mansfield, UK

Lizzie Sagoo
ADAS
Sustainable Agricultural Systems
Cambridge, UK

Roger Robson-Williams
Plant and Food Research
Canterbury Agriculture and Science Centre
Lincoln, New Zealand

ISSN 2511-2260 ISSN 2511-2279 (electronic)
Progress in Precision Agriculture
ISBN 978-3-031-96533-3 ISBN 978-3-031-96534-0 (eBook)
https://doi.org/10.1007/978-3-031-96534-0

This Springer imprint is published by the registered company Springer Nature Switzerland AG
The registered company address is: Gewerbestrasse 11, 6330 Cham, Switzerland

Foreword

The application of technology to horticulture is revolutionising the sector. As the Chief Executive Officer of AgriTech New Zealand I have the privilege to witness first hand some of the global innovation pipeline that is transforming the way we grow, harvest and store horticultural produce. Many of the leading scientists, agronomists, engineers and extension specialists involved in that pipeline are contributors to this book.

Addressing the challenges of climate change, biodiversity loss and pollution while producing healthy fruit and vegetables to feed a growing population is a global challenge. It is no coincidence that the chapter authors have been drawn from both northern and southern hemispheres and from regions of horticultural production with widely varying soils and climates. It also reflects the diverse interests and professional networks of the book's editors.

They share my hope that one outcome from assembling the diverse perspectives and expertise represented in this material will be the emergence of new collaborations within the innovation ecosystem reflected in these pages. In turn, those collaborations will lead to more innovative and effective solutions that are well suited to the needs of growers and the industry as a whole.

Drawing from experts in annual and perennial horticulture, soils, irrigation, robotics, remote sensing and data science, the chapters of this book traverse the application of precision horticulture to every aspect of crop management. Starting with an understanding of the importance of soil variability in fields, orchards and vineyards and concluding with the role of data science in providing a feedback loop, this book is an authoritative summary of the role that precision technology can play in the production of field, orchard and vineyard horticultural crop production. It is an invaluable reference point for my work as a leader of an agricultural technology industry association. I believe it will be equally indispensable to researchers, policymakers, post-graduate students, horticultural advisors and growers as they seek a better understanding of the challenges and opportunities facing the industry.

AgriTech New Zealand, Auckland, New Zealand
March 2025

Brendan O'Connell

Competing Interests The editors have no conflicts of interest to declare that are relevant to the content of this book.

Acknowledgements

The editors are grateful to the chapter authors for their patience in the preparation of the manuscript for this book, which commenced in 2021 and was delayed by the impacts of the global Covid-19 pandemic.

About the Book

This book provides an authoritative summary of the role that precision agriculture plays in the production of field horticultural crop production and how science and engineering could redefine its place in our fields, orchards and vineyards. Bringing together an overview of the current state of the art, case studies of real-life applications of agri-technology in horticulture and evidence-based perspectives on likely future developments, the book includes chapters addressing the challenges, limitations and potential of precision farming technologies in the production, harvesting and post-harvest sorting of a range of annual and perennial horticultural crops. The provision of fresh produce presents unique challenges and opportunities amenable to the application of agri-technology to address issues ranging from high labour costs through to food safety and traceability. Indeed, the possibilities presented by automation are catalysing the complete redesign of some field horticultural production systems.

Contents

Abbreviations

3D	Three-dimensional
AI	Artificial intelligence
ANN	Artificial neural networks
AWC	Available water-holding capacity
BC	Before Christ
CA	Cluster analysis
CBR	Case–based reasoning
CDA	Canonical discriminant analysis
CE	Common era or Christian era, i.e. after the birth of Christ
CIGreen	Chlorophyll index green
CNN	Convolutional neural network
CT	Computer tomography
CTF	Controlled traffic farming
CWSI	Crop water stress index
DA	Discriminant analysis
DAPS	Decision algorithm for patch spraying
DMC	Dry matter content
DSM	Digital soil mapping
DSS	Decision support system
EC	Electrical conductivity
ECa	Apparent electrical conductivity
EMI	Electromagnetic induction
EPN	Entomopathogenic nematodes
ER	Electrical resistivity
Etc	Evapotranspiration
Faster R-CNN	Faster regional convolutional neural network
FOPS	Future orchard planting systems
GHG	Greenhouse gas
GIS	Geographic information system
GNSS	Global Navigation Satellite System

GPR	Ground penetrating radar
GPS	Global positioning system
GSD	Ground sampling distance
GUI	Graphical user interface
HCP	Horizontal co-planar (receiver arrays)
HOG	Histogram of oriented gradients
ICT	Information and Communications Technologies
IoT	Internet of things
IPM	Integrated pest management
IRT	Infrared thermometry
IWM	Integrated weed management
Kc	Crop coefficient
LAI	Leaf area index
LBP	Local binary patterns
LED	Light-emitting diode
LiDAR	Light detection and ranging
LoRa	Long range (networks)
MTCI	Meris Terrestrial Chlorophyll Index
MCARI2	Modified Chlorophyll Absorption in Reflectance Index 2
ML	Maximum likelihood
MRI	Magnetic resonance imaging
MTS	Minimum taste standards
NDVI	Normalised difference vegetation index
NIR	Near-infrared
NIRS	Near infrared spectroscopy
NMR	Nuclear magnetic resonance
NNA	Neural network algorithm
OBIA	Object-based image analysis
OCT	Optical coherence tomography
PA	Precision agriculture
PCA	Principal component analysis
PDU	Plant discrimination unit
PI	Precision irrigation
PRP	Perpendicular (receiver arrays)
PS	Proximal sensors
PWM	Precision weed management
RF	Random forest
RGB	Red, green and blue
RS	Remote sensors
RTF	Random traffic farming
RTK	Real-time kinematic
SAR	Synthetic aperture radar
SBI	Soil brightness index
sCTF	Seasonal controlled traffic farming

SIMCA	Soft Independent Modelling of Class Analogy
SOC	Soil organic carbon
SOM	Soil organic matter
SPA	Successive projection algorithm
SSC	Soluble solids content
SSD	Single shot multibox detector
SSWM	Site-specific weed management
SVM	Support vector machines
SWB	Soil water balance
SWC	Soil water content
SWNIR	Short wave near infrared
Ta	Air temperature
Tc	Canopy temperature
TLS	Terrestrial laser scanner
TZG	Tate ZespriTM grade
UAV	Unmanned aerial vehicle
USD	United States dollar
UV	Ultra-violet
UVE	Uninformative variables
Vis-NIR	Visible to NIR
VNIR	Visible to NIR
VOCs	Volatile organic compounds
VRHS	Variable-rate herbicide spraying
VRI	Variable rate irrigation
WSN	Wireless sensor networks
WUE	Water use efficiency
XRF	X-ray fluorescence
YOLO	You only look once

Chapter 1
Introduction

Paul Newell Price, Lizzie Sagoo, and Roger Robson-Williams

Abstract Humanity faces significant challenges over the coming decades, including how to feed a healthy diet to a growing population without further impacting on earth's essential life supporting systems. Climate change mitigation and adaptation; the need to combat the biodiversity crisis; and increasing pressure on land, soil, air and water will necessitate a move towards an increasingly plant-based diet and more efficient production systems including improved irrigation and nutrient use. Precision technologies could have an important role to play in this transition. This book provides an overview of the current state of the art, case studies of real-life applications of agri-technology in horticulture, and evidence-based perspectives on likely future developments. It addresses the challenges, limitations and potential of precision farming technologies in the production, harvesting and post-harvest sorting of a range of annual and perennial horticultural crops. It is hoped that this book will help readers understand the principles of precision in horticulture production, the potential these technologies hold to transform the industry, and some of the limitations to realising that potential.

Keywords Precision horticulture · Planetary boundaries · Climate change · Crop inputs · Agri-technology · Automation · Robotics

Humanity faces significant challenges over the coming decades. Industrial and agricultural activities have depleted finite resources and accumulated pollutants to the extent that six out of nine planetary boundaries (defining the safe operating space for the earth's essential life supporting systems) have been exceeded (Richardson

P. Newell Price (✉)
ADAS Gleadthorpe, Mansfield, UK
e-mail: Paul.Newell-Price@adas.co.uk

L. Sagoo
ADAS Boxworth, Cambridge, UK

R. Robson-Williams
Plant & Food Research, Canterbury Agriculture & Science Centre, Lincoln, New Zealand

P. Newell Price et al. (eds.), *Precision Agriculture in Field Horticulture*, Progress in Precision Agriculture, https://doi.org/10.1007/978-3-031-96534-0_1

et al. 2023). Food production is recognized as a major driver of planetary boundary transgressions (Campbell et al. 2017) but progress on globally agreed goals for alleviating hunger is not proceeding at required or agreed rates (UN 2023). The challenge of feeding a healthy diet to a growing population is almost certain to become more intractable as food production is increasingly impacted by the biosphere instability caused by human activity.

Specifically, higher temperatures and shifting weather patterns will alter land capability, with some regions declining in productivity while others potentially increase (Shukla et al. 2019). Climate change mitigation and adaptation will necessitate a move towards an increasingly plant-based diet with lower meat and milk production per capita. There is also an urgent need to combat the biodiversity crisis and to conserve at least 30% of the globe as natural ecosystems. The pressure on land, soil, air and water will also require increasingly efficient production systems including improved irrigation and nutrient use. These changes will help meet several of the United Nations' Sustainable Development Goals, contributing towards a future sustainable food system. To improve human health and reduce demands on medical services, our increasingly plant-based diet will also need to contain a higher proportion of pulses, fruit, nuts and vegetables. Growing horticultural crops in an efficient manner will therefore become increasingly relevant.

Horticultural production itself is changing and has its own challenges. Labour shortages are widespread and costs of production increasing, with the price of fertiliser and fuel fluctuating dramatically in recent years. This has increased the demand for greater efficiency and automation alongside production system changes such as trellising, training and 'two-dimensional' canopy systems for top fruit. There may also be the opportunity to produce annual and perennial horticultural crops close to urban centres on land previously used for arable or livestock production. In many instances, this will require a re-evaluation of soil and land suitability to optimise system efficiency.

There is an increasing need to manage soil, land, crops and inputs with greater precision to optimise yields, improve efficiency, conserve resources and reduce environmental impact of agricultural and horticultural production. Precision agriculture (PA) can be defined as management of spatial and temporal variability in fields and crops using information and communications technologies (ICT) (Blackmore et al. 2003). It has potentially wide application in horticultural systems. This book sets out the potential for precision approaches and technologies to facilitate innovation and drive value in outdoor fruit and vegetable production. Bringing together examples of the current state of the art, case studies of real-life applications of precision technologies in horticulture, and evidence-based perspectives on likely future developments, the book includes chapters addressing agritechnology in the production, harvesting and post-harvest sorting of a range of annual and perennial horticultural crops. The provision of fresh produce presents unique challenges and opportunities amenable to the application of agri-technology to address issues ranging from high labour costs through to food safety and traceability. Indeed, the possibilities presented by automation are catalyzing the complete redesign of some horticultural production systems.

The efficiency of outdoor fruit and vegetable production is determined to a significant extent by climate and soil suitability. It is also important to know how soil conditions and soil nutrient supply vary between and within fields so that cultivations can be timed appropriately, and lime, fertiliser and water used efficiently. Chapter two therefore covers soil sampling and sensing to improve understanding and management of soil variation within horticulture systems. The chapter also investigates the barriers to further adoption of soil sensing, sampling and survey technologies, many of which are common to other precision horticulture technologies.

The implementation of agri-technology within horticultural systems requires accurate positioning and guidance of machinery. This is covered in detail by Heege (2013) so is not repeated in this book, although the basic principles are covered here. Machinery positioning and guidance in horticultural systems is principally achieved by satellite-based navigation systems, generically referred to as the Global Navigation Satellite System (GNSS). The degree of positioning accuracy increases with the number of satellite signals that are available to the receiver and the use of a correction signal that can be provided by an orbiting satellite (lower accuracy) or a static terrestrial reference point, e.g., a local base station (greater accuracy). Unenhanced GNSS locates the user to within around 1 m but is not sufficiently accurate for most horticultural operations. Differential GNSS (DGNSS) uses a minimum of five satellites and the position of a fixed object to correct for GNSS drift (change in the calculated position by up to 1.5 m per hour due to satellite orbit errors and fluctuating weather conditions) and can achieve sub 1 m pass-to-pass accuracy to within +/− 10 cm. However, many automated horticultural operations require Real Time Kinematic (RTK) systems that use a local base station. This comprises a DGNSS receiver and a radio transmitter, to provide a correction signal; and can achieve pass-to-pass and static accuracies of +/− 1–2 cm.

Many horticultural production systems involve multiple machinery passes to carry out operations such as cultivating, drilling, planting, weeding, fertilising and harvesting. If machinery movements and timings are not controlled and operations are carried out in wet field conditions, this can result in significant soil damage (e.g., soil compaction, smearing and 'puddling'), which compromises the efficiency of subsequent nutrient and water use, and of production as a whole. Harvesting by hand in combination with machinery can have similar effects. Chapter three therefore reflects on the role of Controlled Traffic Farming (CTF) in horticultural systems and its potential to reduce costs and environmental impacts while potentially increasing yields and efficiency of production.

Horticultural production requires the control of weeds, pests and diseases, and the provision of water and nutrients to meet crop requirements. Chapter four describes the precision management of pests and diseases in horticultural systems, with a focus on the application of pest forecasts and their role in precision horticulture. It provides an overview of the options for precision detection tools currently available and how they facilitate precision management. Chapter five reviews the use of sensors to identify and control weeds in horticulture. It includes case studies describing how sensing technologies are used to guide mechanical weed control and the precise application of herbicides, and discusses future challenges to the increased adoption of precision weed management. Chapter six describes the latest *state-of-the-art* technologies that enable precision irrigation management in outdoor horticulture systems. It provides a review of sensing technologies and their integration into precision irrigation decision support systems, including variable rate delivery technologies to optimise water, energy consumption and other environmental impacts, and identification of current research in precision irrigation application for horticulture crops, including barriers and future research in precision irrigation. Chapter seven covers the economic and environmental importance of accurate crop nutrient management in outdoor horticultural systems. It describes precision nutrient management based on the measurement and management of soil and crop properties, focusing on case studies in vegetable production.

In the growing and harvesting phases of horticulture, production precision technology can be used to support yield forecasting and yield mapping. Accurate pre-harvest yield forecasting can improve the efficiency of the harvest operation, by allowing growers to better schedule harvest and provide accurate information on harvest volumes to processing factories and/or buyers. Yield mapping at harvest has potential to provide useful information about the spatial variability of yield and thereby inform growers on how input management and economic performance can be improved, and environmental impacts reduced. A review by Longchamps et al. (2022) has very adequately summarised the current *state-of-the-art* for yield mapping in horticulture and so is not repeated in this book. The paper explains the basic principles behind proximal (direct and indirect) and remote yield monitoring approaches as a way to estimate and forecast yield prior to harvest, and provides examples of application in horticultural cropping systems.

For many horticultural producers around the world, finding sufficient manual labour at reasonable cost for planting, weeding, pollination, thinning (of buds, flowers or fruitlets), harvesting and sorting is becoming increasingly challenging. There is considerable potential for robotics to address such labour shortages, reduce work-related injuries, increase productivity and minimise waste. However, there are also considerable technical challenges to be overcome. Chapter eight takes a detailed look at pollination in kiwifruit, and harvesting in apples and kiwifruit, to explore some of the most intractable problems in automating horticultural production. A case study of harvesting a horticultural row crop is also provided.

Post-harvest processing of horticulture produce requires sorting and grading of fruit and vegetables in terms of size and quality. Chapter nine provides an overview of modern sorting and grading technologies, with a focus on the commercial use of machine vision and near-infrared spectroscopy (NIRS) and the development of firmness grading methods for fruit. Prospects for future innovations are discussed, including advances in data processing algorithms and the deep imaging methods of MRI and X-ray technology. Barriers to adoption are considered and two case studies are presented illustrating the opportunities and problems encountered with commercialisation.

The use of precision technologies results in copious amounts of data, which can be captured and exploited in several ways to benefit a horticultural enterprise. Methods of data collection and analytics are rapidly evolving with recent advances in real-time or near-time sensing, monitoring, communication and data storage solutions. Demand for integrated data solutions is rapidly increasing due to the value they can bring to the horticulture industry. This in turn is raising issues around trust, licensing and data sharing that need to be resolved. Chapter ten reviews the phases that must be undertaken carefully in data-driven horticultural studies to allow the reuse and sharing of data in the future. It considers the interaction between horticultural and modern data science and explains how data science can act as a force-multiplier for horticultural research and shape new directions and aims in this field.

Precision technologies have multiple applications in horticulture and the potential to increase production efficiency in outdoor production systems. However, challenges remain in overcoming barriers to their adoption through the provision of sufficient independent technical and financial support to growers. It is also important to consider when and where the adoption of such technologies is appropriate, depending on the scale of operation, the mindset of the grower and the amount of technical support available.

A further consideration is where any benefits from the application of precision horticultural practices accrue. Agriculture and horticulture have been responding to environmental concerns for decades with incremental efficiency improvements. Progress has been made in increasing crop yields and optimising inputs of e.g., water and fertiliser. However, such efficiency gains tend to reduce costs, leading to increased demand and thereby resulting in increased rather than decreased resource use. This is a phenomenon known as the Jevons' Paradox (Polimeni and Polimeni 2006). The pursuit of efficiency alone is unlikely to be sufficient to address the dilemma of providing healthy diets for a growing population using food production systems that operate within ecological limits.

It is hoped that this book will help readers understand the principles of precision in horticulture production, the potential these technologies hold to transform the industry, and some of the limitations to realising that potential.

Competing Interests The authors have no conflicts of interest to declare that are relevant to the content of this chapter.

References

Blackmore S, Godwin R, Fountas S (2003) The analysis of spatial and temporal trends in yield map data over six years. Biosyst Eng 84:455–466

Campbell BM, Beare DJ, Bennett EM, Hall-Spencer JM, Ingram JSI, Jaramillo F, Ortiz R, Ramankutty N, Sayer JA, Shindell D (2017) Agriculture production as a major driver of the earth system exceeding planetary boundaries. Ecol Soc 22(4):8. https://doi.org/10.5751/ES-09595-220408

Heege HJ (2013) Precision in guidance of farm machinery. In: Heege H (ed) Precision in crop farming. Springer, Dordrecht. https://doi.org/10.1007/978-94-007-6760-7_4

Longchamps L, Tisseyre B, Taylor J et al (2022) Yield sensing technologies for perennial and annual horticultural crops: a review. Precision Agric 23:2407–2448. https://doi.org/10.1007/s11119-022-09906-2

Polimeni JM, Polimeni RI (2006) Jevons' paradox and the myth of technological liberation. Ecol Complex 3:344–353

Richardson K, Steffen W, Lucht W, Bendtsen J, Cornell SE, Donges JF, Drüke M, Fetzer I, Bala G, von Bloh W, Feulner G, Fiedler S, Gerten D, Gleeson T, Hofmann M, Huiskamp W, Kummu M, Mohan C, Nogués-Bravo D, Petri S, Porkka M, Rahmstorf S, Schaphoff S, Thonicke K, Tobian A, Virkki V, Wang-Erlandsson L, Weber L, Rockström J (2023) Earth beyond six of nine planetary boundaries. Sci Adv 9(37):eadh2458

Shukla PR, Skea J, Slade R, van Diemen R, Haughey E, Malley J, Pathak M, Portugal Pereira J (eds) (2019) Technical summary. In: Climate change and land: an IPCC special report on climate change, desertification, land degradation, sustainable land management, food security, and greenhouse gas fluxes in terrestrial ecosystems https://doi.org/10.1017/9781009157988.002

United Nations (2023) https://sdgs.un.org/sites/default/files/2023-07/The-Sustainable-Development-Goals-Report-2023.pdf. Accessed 10 Aug 2025

Chapter 2
Soil Sensing and Sampling

Paul Newell Price, Timo Samuel Breure, Lizzie Sagoo, and Jack Hannam

Abstract A good understanding of soil spatial variability within and between fields can help improve productivity in terms of overall crop yield, quality and evenness; and the use of nutrients and appropriate cultivations. This chapter sets out the main options for determining the variability of soil properties and how this can be developed into effective soil and land management strategies. There are a range of field survey approaches, technologies and statistical methods that can be used in isolation or in combination to map variation in soil properties; from soil samples analysed in the laboratory, soil survey (using a spade, auger and an experienced soil surveyor) to proximal sensing (e.g. electromagnetic induction) and geostatistical survey. The most appropriate and cost-effective method will depend on the amount of prior information on soil variability, the local cost of available skills and technologies (e.g. the cost of soil analysis), the scale of operation, and the specific objectives of the survey (i.e. how does the grower intend to use the spatial information to vary their land management?). Where strong spatial variability in soil physical properties is expected (e.g. variability in sand, silt, clay or organic matter content) but there is no prior spatial information on that variability, the use of proximal sensing may be useful as a means of guiding soil sampling schemes and establishing soil management zones. Geostatistical surveys are especially useful where weak spatial variability is expected in soil physical properties (e.g. soils formed from widespread uniform sandy or silty deposits) but strong spatial variability is expected in soil chemical properties, such as soil nutrient reserves, which may have arisen due to contrasting management between or within fields. In this context, a spatial covering design with

P. Newell Price (✉) · L. Sagoo
RSK ADAS Ltd, Helsby, UK
e-mail: Paul.Newell-Price@adas.co.uk

T. S. Breure
Rothamsted Research, Harpenden, UK

Cranfield University, Cranfield, UK

J. Hannam
Cranfield University, Cranfield, UK

P. Newell Price et al. (eds.), *Precision Agriculture in Field Horticulture*, Progress in Precision Agriculture, https://doi.org/10.1007/978-3-031-96534-0_2

additional close pair points can yield useful information but the overall costs of sampling and analysis should be borne in mind.

Keywords Soil variability · Soil sampling · Soil sensing · Soil analysis · Soil management · Geostatistics · Sampling costs

2.1 Introduction to Soil variability and Effects on Crop Production Efficiency

The management of soils and their properties is critical to the efficiency of crop production. Soils have multiple properties. Some are inherent and cannot be changed readily, and others are dynamic and can be changed through management. Inherent properties include soil texture, soil depth and stoniness. Dynamic properties include soil pH, nutrient reserves, soil organic matter (SOM) and soil biology. Inherent and dynamic soil properties vary in three dimensions, across the field or landscape and vertically with depth by soil layer or horizon. Soil management, land use and crop type often follow soil suitability and land capability, determined by inherent soil properties. Management can be further used to optimise dynamic properties for horticultural production, for example by adjusting soil pH with the addition of lime. Soil properties determine how soils and crops respond to cultivation, irrigation, and the application of lime and nutrients. Optimal production efficiency relies on timely and targeted cultivation and applying the right quantities of lime, nutrients and water when and where crops most need them. To do this effectively it is important to have a good understanding of soil variability, so that management can be tailored as soil properties change across a farm or nursery, and even within a field.

There are several factors that influence how soils form, resulting in the spatial variation of soil types and properties. At a global scale, the 'CLORPT' formula, defined by Jenny (1941), describes soil formation as a function of CL = climate, O = organisms, R = relief, P = parent material, and T = time. However, at the farm or field scale the primary factors that determine soil variability are 'organisms' (land and cropping history), relief (slope) and parent material (the geological or biological matter from which soils have formed). A change in soil parent material or slope across a farm or field can cause a change in soil type or properties. In addition, some factors such as soil parent material can have different degrees of variability. For example, soils formed from the same geological material or from sediments deposited by water, ice or wind can be relatively uniform at the field scale. This is often relevant to sediments formed on the bed of a lake (lacustrine sediments) or 'drift' from the melting of glaciers or wind-blown silts. However, some parent materials can exhibit mixtures of different particle sizes, producing soils that are spatially variable over short distances. Parent materials such as alluvial materials deposited by rivers, mixed 'glacio-fluvial' deposits from outwash of glacial lakes or colluvial materials at the base of a hillslope, are often very variable (e.g. Fig. 2.1).

Fig. 2.1 Contrasting soil variation between fields in the Cambridgeshire Fens, England, where the peat cover has wasted to reveal a dendritic pattern of silty 'rodhams', which mark the position of former creeks and give a complex soil pattern with silty soils on ridges and clayey and peaty soils on intervening ground. Imagery taken from Google Satellite 2023 (Imagery ©2023 Airbus, Landsat / Copernicus, Maxar Technologies, Map data ©2023). The coordinate references (EPSG: 4326–WGS 84) of the tile in longitude/latitude are: −0.211, 52.513)

Soil variability across a field can affect optimal crop management in several ways impacting the timing, type and ideal depth of cultivations; crop suitability; available water supply (and the related need for irrigation); potential yield; the quantity of major nutrients needed; and the likelihood of trace element deficiencies. Soil drainage properties will determine if a field needs an underdrainage system, how often the land can be accessed (without causing severe compaction) and how rapidly it warms up in the spring. Topsoil texture is an important factor determining how the soil reserves of major nutrients, such as phosphorus (P) and potassium (K), should be managed. 'Clayey' soils (by definition) contain a high proportion of clay mineral particles with a negative charge that contributes to cation exchange capacity. They can hold on to and in some cases supply potassium through weathering, but their high sorption capacity means that P availability can be naturally low. By contrast, sandy soils have a low clay mineral content and tend to have low cation exchange capacity and P sorption capacity. Their ability to hold on to and supply K is therefore often low, while P availability can be higher than in soils with higher clay content.

Due to the contrasting nature of 'sandy' and 'clayey' soils and the different ways they respond to management, it can make good practical sense to manage them separately. Growers aim to cultivate land when the soil at cultivation depth is in a 'friable' state, so that it shears rather than smears or compacts as an implement is pulled through it. In any given season, 'sandy' soils dry out more rapidly than 'clayey' soils and so two contrasting soil types can often require different cultivation depth and timing to create a good seedbed and soil structure, and to avoid structural damage that can inhibit crop growth. Contrasting soil types also require a different frequency and quantity of lime and nutrient applications. For example, a medium-textured calcareous soil with moderate P and K reserves has no lime requirement but could need regular applications of phosphate and potash. By contrast, a light-textured, naturally acidic soil (low clay content) with high P reserves and low K would need regular lime and potash applications, but limited phosphate. Managing these two soil types in the same way would lead to low nutrient use efficiency and significant limitations to crop yield and quality.

This spatial variability of soil is itself highly variable between agricultural fields and, for some management considerations such as nutrient management, can determine the relevance of a within-field precision approach to horticulture. In these contexts, the specific management needed to optimise horticulture production can be very different from one part of a field or farm to another. Optimising crop yield means managing the growth limiting factors that characterise each area of land. By varying soil and crop management to take account of spatial variation in the principal growth limiting factors, production efficiency can be optimised. The key to practical management is deciding whether the variation in soil properties and growth limiting factors within a field or farm warrants such spatial adjustments to soil and crop management.

Growers that have managed a plot of land for many years will know how soil and other growth factors vary across it. Crop rotations, cultivations and sometimes nutrient applications may be adjusted to account for this variation. However, for new fields, nurseries and farms, how should a grower determine their variability? In many instances, no soil maps or prior spatial soil sampling or surveying will be available. There are two common approaches to assessing or mapping soil variability within an agronomic context. One relies on a combination of soil sampling, soil analysis and mathematical interpolation to produce a map of how soil properties vary across a field; this technique is reliant on the accuracy of the values derived from soil sampling and analysis to characterise soil variation through geostatistics. The other approach uses a combination of soil sensing, soil survey and soil sampling and analysis. Firstly, soil sensing or soil survey (or a combination of the two) are used to determine the main soil types on an area of land and the location of the boundaries or intergrades between the soil types. Secondly, the soils within each land unit are sampled and analysed to determine their properties. The following section describes the techniques and technologies that can be used to determine soil spatial variability.

2.2 Methods to Assess Soil Spatial Variability

Existing soil survey maps and local farmer knowledge can provide detailed information about soil variability. However, even in countries where an extensive soil survey has been carried out, only a small proportion of soils (e.g. 25% in England and Wales) have been mapped at sufficient scale (e.g. 1:25,000) to determine within-field soil type variation (King et al. 2003). The traditional method involves taking soil samples in the field and characterising these in the laboratory. Nowadays, there are a variety of proximal- and remote sensor technologies, such as electromagnetic induction and soil brightness imagery that can help growers understand and map soil variability effectively and efficiently. Other such methods include infrared and x-ray fluorescence spectroscopy. For all sensor technologies, calibration to laboratory measurements is required to get a quantitative estimate of a soil property. The number of calibration samples required depends on the sensor characteristics, soil variation within the site of interest, the sampling scheme and the desired accuracy to estimate the soil property under study.

2.3 Geophysical Ground Survey

Soil Electrical Conductivity

Soil electrical conductivity (EC) is a measure of the soil's ability to conduct an electric current and can reflect differences in soil texture, soil moisture and organic matter content. There are currently two main types of commercially available sensors that differ in the configuration of how electrical conductivity or resistivity is measured: non-contact Electromagnetic Induction (EMI) scanners and 'contact' soil EC scanners:

- EMI sensors are held above the soil when scanning (Fig. 2.2). The sensors have a transmitter and receiver coil at opposite ends of the unit and a sensor in the device

Fig. 2.2 Soil EMI scanner

Fig. 2.3 Soil EC scanner

measures the resulting electromagnetic field that the current induces to provide a measure of the apparent electrical conductivity (EC) of the soil. The strength of the secondary electromagnetic field is proportional to the soil EC.

- Soil EC scanners make contact with the soil when scanning and measure variation in the EC of the soil (Fig. 2.3). Typically, two or three pairs of coulters are mounted on a toolbar often in a Wenner array; one pair provides an electric current into the soil (transmitting electrodes) and the other coulters (receiving electrodes) measure the voltage drop between them.

Although the scanners use different methods of measuring soil EC and the magnitude of measured EC may differ, the patterns of spatial variation are comparable. Hence, the measured variation in soil EC tends to be strongly correlated when performed on the same field (King et al. 2003; Sudduth et al. 1998, 2003, 2005; Heege 2013).

EMI and EC surveys are normally conducted when the soil is bare after one crop is harvested and before the next crop is planted. The instruments are pulled across the field typically at widths of 12–24 m and the information is combined with global-positional data (GPS) to produce a soil EC map. Good soil-to-coulter contact is needed for contact EC sensors and therefore scans are usually carried out following harvest prior to cultivation. Soil EC measurements should not be taken when the soil is frozen as this can affect the measurement.

Most EMI and EC scanners will measure conductivity for two depths of soil simultaneously, providing EC maps for a shallow and deep vertical cross section of soil. Shallow EC measurements are most effectively used for directing soil sampling, while deeper EC measurements can be used to indicate crop yield potential and to compare soil EC and crop yield maps. The depths vary depending on the configuration of the transmit and receiver coils, but many report measurements at 40 cm and 120 cm.

The main factors affecting soil EC are soil texture, moisture content, organic matter content, salt content, temperature and soil bulk density. In most fields, soil texture is the main cause of soil EC variation. Clay soils with high particle-to-particle contact and high moisture holding capacity are highly conductive, whilst sandy soils with limited particle-to-particle contact and low moisture holding capacity are poor conductors. Soil moisture content may affect the magnitude of EC values measured, but should not affect the pattern of variability—a soil EC map will consistently identify areas of different soil texture regardless of the soil moisture content at the time of measurement, although the boundaries between contrasting soil types may change slightly between sampling times. Grisso et al. (2009) found that fields mapped several times during the year at differing soil moisture contents had a varying magnitude of EC values but consistently identified approximately the same pattern of variation in soil texture. Soil texture is an inherent soil property (i.e. it does not change over time), so soil EC mapping only needs to be done once.

Given the suite of soil properties that affect EC measurements (soil texture, moisture content, organic matter content, salt content, temperature and soil bulk density), which can be expected to correlate, it is difficult to separate individual effects. In some circumstances, it may be possible to attribute changes in EC measurements to a specific soil property, such as the detection of soil compaction (Besson et al. 2004; Krajčo 2007; Galambošová et al. 2020). However, this depends on repeat EC measurements carried out at a similar soil moisture content and severe compaction to depth. Soil compaction is rarely visible on an EC map, as a compacted soil layer often represents only a small percentage of the assessed soil volume.

Ground Penetrating Radar

In Ground Penetrating Radar (GPR), the transmitting antenna generates a broadband electromagnetic signal that propagates through the soil and is reflected at boundaries of materials with differing dielectric properties, i.e. the ability of a material to store or dissipate an electrical or magnetic charge. The dielectric property of the soil affects the velocity and reflectivity of the electromagnetic wave. Heterogeneities in soil characteristics in 2, 3 and 4 dimensions can also give rise to differing dielectric properties. The largest effect on GPR is the dielectric constant or relative permittivity, which is a material's ability to hold or oppose an electric charge. In soil, relative permittivity is primarily influenced by water, given the large permittivity of water compared to other soil materials. The depth that the electromagnetic signal can penetrate the soil is affected by the conductivity of the medium. GPR can be effectively used to several tens of metres depth in sandy soils with low conductivity. However, in soils with higher conductivity, such as very saline soil or those with high clay contents, the depth penetration is limited and therefore information on the subsurface can be constrained.

Due to the primary influence of water on permittivity, GPR has been effectively used to map soil water contents at different depths (Lambot et al. 2009; Huisman et al. 2003; Lombardi et al. 2022). Other applications of GPR for mapping soil properties include soil texture (Boll et al. 1996; Sarkar et al. 2019), soil compaction (Zajícová and Chuman 2019), soil salinity, water table and soil depth (see Chap. 6 on

"Precision Irrigation in Outdoor Horticulture"). These applications capitalise on the contrasts of dielectric properties between different materials such as soil texture, bulk density of compacted layers and contrasting textures of different soil horizons. GPR can also potentially be used to detect spatial differences in rooting depth. In a review, Liu et al. (2016) stated that GPR holds the potential for accurate in-situ measurements of root phenology within the context of agriculture. Although the physical principles of the effects of roots on soil dielectric permittivity are unclear, the water potential held by the root network should be captured by GPR on a bulk soil volume scale (Liu et al. 2016). The latter being so because the difference in the dielectric constant between water, air and minerals is quite distinctive (Viscarra Rossel et al. 2011).

2.4 Spectroscopy

Infrared and X-Ray Fluorescence Spectroscopy

The main principle of soil spectroscopy is that electromagnetic radiation will induce an interaction with the soil sample or bare ground surface to be monitored (such as absorption, fluorescence, emission or diffraction). The most common response of interest for soil spectroscopy is absorption at different wavelengths across the visible and infrared spectrum that produces a unique spectral signature. Measurements can be taken by benchtop-, handheld- or remote spectrometers (such as satellite imagery).

In soil sensing, a spectrometer emits electromagnetic radiation directly onto a discrete soil sample or spot of bare ground in the field, and the response from the soil at different wavelengths corresponds to excitation at a molecular level of different soil components (such as the bonds between different elements in molecules). With 'active' sensing technologies, the radiation source is emitted by the sensor itself, as is the case with Synthetic Aperture Radar (SAR), whereas 'passive' sensing technologies use the sun as a radiation source, such as the Landsat satellites that produce multispectral imagery.

Application of remotely sensed multispectral imagery of bare soil has been used to map soil variability using the soil brightness index (SBI). The brightness index is based on the red and green multispectral bands:

$$\mathrm{SBI} = \sqrt{\frac{R^2 + G^2}{2.}}$$

Where SBI is the soil brightness index, R is the Red band (0.64–0.67 μm) and G is the Green band (0.53–0.59 μm).

SBI is responsive to changes in soil colour (Escadafal 1989), which is related to changes in soil moisture, soil mineralogy (e.g. iron oxides), organic matter content and light-coloured materials (e.g. gypsum, salt crusts, carbonates). Thus, SBI has been used to map saline soils (Allbed and Kumar 2013), soil organic carbon (Gholizadeh et al. 2018) and iron oxides in soil (White et al. 1997).

Soil infrared spectroscopy determines the fraction of radiation that is absorbed at a particular energy, representing fundamental vibrations of atoms in both organic and inorganic molecules (Du et al. 2015; Stenberg and Viscarra Rossel 2010). Although the principle is the same across the visible, near- and mid-infrared light ranges, their specific application differs in the characteristic of vibrations detected and how these correspond to soil properties. For example, near-infrared is mainly characterised by vibrations caused from electronic excitations mostly associated with minerals containing iron. Absorption bands within the near-infrared often overlap as well, making them less suitable for a qualitative analysis compared to mid-infrared. The mid-infrared range in its turn is characterised by fundamental vibrations (no overlap) of soil components such as soil organic carbon fractions or quartz (Stenberg and Viscarra Rossel 2010).

X-ray fluorescence (XRF) spectroscopy induces the translocation of electrons within atoms present in the soil. The energy intensity emitted during these translocations (i.e. the fluorescence) is measured over a set time period. Each wavelength in the X-ray region is associated with a given element (although overlaps do occur) and thus the energy intensity measured can be related to the total content of elements in the soil. XRF spectroscopy has had limited applications within agriculture and horticulture relating to soil, where research covers mainly the topic of detecting heavy metal contamination in either the soil or harvested products (Weindorf et al. 2012).

Given the principles of spectroscopy, spectra are in essence a proxy of the soil variable of interest such as, for example, organic carbon. Thus, to quantitatively predict organic carbon content, spectra have to be calibrated by relating them to a known (sub)set of reference samples through an empirical relation (i.e. regression). The (sub)set of known reference samples is required to be 'representative' of the target set for which organic carbon will be predicted. The concept of a 'representative' sample can take on several meanings, such as the coverage of geographical/pedological/spectral variation. An optimal sample for calibration purposes is 'representative' such that it covers the spread of the target variable. In a sense, it might be biased when compared to the target population mean for which the calibration model is used to predict. The procedure of using a representative subset to predict soil physical (e.g. particle size distribution), chemical (e.g. major nutrients) and biological (e.g. microbial C) properties based on infrared or X-ray fluorescence spectroscopy has been reported extensively within the scientific literature (Kalnicky and Singhvi 2001; Sinfield et al. 2010; Kuang et al. 2012).

The spectral signature can also be affected by irrigation and cultivation timing. This should therefore be taken into account in any calibration. If appropriately calibrated, soil spectroscopy is cost-effective when compared to traditional laboratory methods. Research grade visible, near-infrared spectrometers cost in the range of USD 50,000 to 60,000, whereas mid-infrared and X-ray fluorescence spectrometer costs range from 60,000 to 100,000 USD and above. However, given that an instrument allows for scanning tens of thousands of samples during its lifetime, the overall price per sample analysed can lie well below that of traditional laboratory methods. Purchase costs vary across the globe as well as with the sensor characteristics, such as its portability, sample through-put and wavelength coverage.

Sensor Fusion

The combination of sensors that are based on different physical principles to predict soil properties is referred to as sensor fusion. Either measurements can be combined prior to the calibration regression or different calibration regressions can be combined in a model ensemble. The general expectation is that combining different sensors will create more robust predictions of soil properties. Also, the confidence in predictions is increased since multiple independent measurements are made on the same soil aliquot or sample. Furthermore, each sensor has a unique measurement support leading to a higher coverage of the soil domain in question (Adamchuk and Viscarra Rossel 2010). For soil spectroscopy, studies that investigated the combinations of visible, near- and mid-infrared have been inconclusive, reporting both limited benefits (Clairotte et al. 2016; Viscarra Rossel et al. 2006) and general improvements (Viscarra Rossel et al. 2006; Knox et al. 2015; Johnson et al. 2019; Ng et al. 2019) compared with using a single band. Other studies examined the combination of either visible and near-infrared, or mid-infrared with XRF spectroscopy (Towett et al. 2015; O'Rourke et al. 2016a, b; Zhang and Hartemink 2020; Benedet et al. 2020; Oldoni et al. 2024). In most cases, the benefits of multiple sensors have been found to differ between soil properties. However, few studies have investigated the combination of visible, near-, mid-infrared and XRF sensors combined (O'Rourke et al. 2016b). It is also worth noting that the majority of the soil spectroscopy literature on sensor fusion examines large-scale soil databases rather than field-scale data sets.

2.5 Approaches to Mapping Spatial Variability

The previous section detailed different methods and technologies that can be used to assess soil variability. The outputs from geophysical survey and remote sensing may already be at sufficient resolution for application of precision agriculture. For example, SBI can be derived from Sentinel-2 and Landsat 8 satellite imagery that has 10 m or 30 m pixel resolution, respectively. Other methods such as proximal application of spectroscopy rely on taking samples and the production of maps through the application of geostatistics or interpolation methods.

Soil Management Zoning

Geophysical and soil brightness are examples of outputs that can identify where there is variability within a field. Measured soil variation can be made useful for a farming context by creating zones that require different management. Two processes are required: (1) zoning areas where different management can be applied and (2) identifying the primary causes of the variability between the zones by taking representative soil samples from each zone. The zoning should relate to areas demonstrating differences in their viable yield and thus require different management. Following Whelan (2018), management zones should partition horticultural fields such that: (a) the variability (of the agronomic property of interest) within management zones is smaller compared to whole field variability, (b) there is a significant difference between zones in terms of the average within-zone variability, (c) there should be a reduction of variability for agronomic variables that have not been used in the process to classify the field into management zones. Overall, management zones should be invariant across time while accurately characterising the observed variations in crop yield.

Interpolation from Point Samples to Assess Spatial Variability

Another method to determine the spatial variability of soil properties is to use soil sampling and analysis combined with interpolation, i.e. predicting values in between measurement locations. The interpolation method involves an assessment of how soil properties vary with distance between sampling points, which involves the calculation of a variogram, a mathematical (or correlation) function that represents the short and large-scale spatial variability of landscape properties. Geostatistics involves model-based predictions by treating the target variable as an outcome of a spatial random process. Geostatistical prediction is done by interpolation methods such as kriging, which involves weighted averaging of the observed values of a property, within a 'neighbourhood' (Webster and Oliver 2007). Kriging allows one to quantify uncertainty at prediction sites. To derive a robust estimate of the correlation function or variogram for a specific field or farm, a minimum of around 50–100 samples has been suggested although this depends on the underlying complexity of the soil variation and its geographical extent (Webster and Oliver 1992; Lark 2000). Furthermore, it involves sampling densities greater than one sample per hectare to estimate short-scale variation.

Within the last 30 years, geostatistics has become more accessible to users due to its integration within existing software packages (Oliver and Webster 2014), which facilitates its use. However, the application of geostatistics requires a detailed check of whether the assumptions of a non-constant mean and variance of the soil property hold true across the site of interest. This is often not the case and thus models have to be generalised to account for a non-constant mean (Lark et al. 2006) or variance

(Marchant et al. 2009) as well as outliers present within the dataset (Marchant et al. 2010). Accounting for deviations from model assumptions thus requires expertise on the subject and complicates geostatistical applications (Arrouays et al. 2012).

Some consideration also needs to be given to the optimal sampling design for geostatistical surveys from a cost effectiveness perspective. It has been established that regularly spaced sampling schemes are inadequate to account for short-range variations in soil properties, which is critical for geostatistical analyses (Starks 1986). However, if regular or grid spacing is not adequate, which scheme is optimal?

Wadoux et al. (2019) contrasted three different sampling schemes with respect to the sample size required to obtain comparable results: (i) spatial coverage sampling schemes (sc); (ii) spatial coverage schemes supplemented with close pairs of points (sc+) and (iii) schemes optimised to reduce the total error. They found that spatial coverage designs with a few added close-pair points (sc+) were almost as accurate as complex optimised designs, and far less costly. Adding a few close-pair sampling points to a spatial coverage scheme improves the accuracy of the derived map without substantially increasing the cost of the survey (Fig. 2.4). Indeed, even if the variogram is known, the benefit of using optimised designs over sc + is generally modest.

Wadoux et al. (2019) concluded that "*there was little evidence of large benefits from optimising sampling schemes. Therefore, it is better in most cases to use a spatial coverage scheme supplemented by a subset of close-pair units unless prior knowledge of the variogram is available (e.g. reconnaissance survey)*". Where the variogram is not available, a field-specific EMI scan may indicate the degree of spatial correlation within a field/farm and can be used to account for a non-constant mean in case it covaries with the soil property of interest. The degree of spatial correlation can indicate whether geostatistical mapping is worthwhile as for example low spatial dependency will require a large number of samples to get a robust estimate of the variogram.

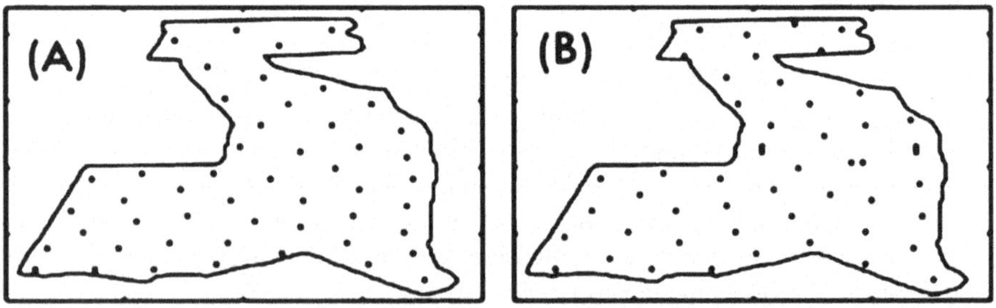

Fig. 2.4 Contrasting sampling schemes, (**A**) and (**B**). A sampling design with close-pair points (**B**) can improve the accuracy of the derived map

Benefits and Applications of Management Based on Soil Sensor Data

The main advantage of soil sensors is that they are rapid, low-cost methods of obtaining high resolution information on within-field soil variability. For example, soil EC maps are provided at a resolution equivalent to the bout width—generally 12–24 m—and, based on a travel speed of 10 km per hour at 24 m bout widths, can cover 19 ha/hour. The main use of soil sensor data is to define soil management zones with similar soil properties, which can be combined for soil sampling and management. Grisso et al. (2009) and Kweon (2012) describe the use of soil EC to identify management zones, a service that is available commercially in many countries.

Soil EC maps are frequently correlated with yield maps reflecting the dominating effect of soil texture and therefore also soil water holding capacity on yield. Where both soil property maps and yield maps are available, this information can be used together to define soil management zones. Soil and yield measurements can also be combined to help establish yield potential maps (e.g. Lund et al. 2000; Marques da Silva et al. 2012). However, most of the soil and yield map comparisons have looked at long season arable crops such as cereals and thus other soil properties are of importance compared to horticulture. That is, the extent to which soil texture and associated soil moisture levels affect yield may be weaker for irrigated horticultural crops. For example, Monaghan (2015) measured lettuce yields (two successive harvests) and soil EC from two fields in Cambridgeshire, England and although the yield patterns were consistent for the two lettuce crops, they did not correlate to soil EC.

The soil zones delineated based on soil sensor data can be sampled separately (for pH, P, K and Mg) to create soil nutrient/pH maps that can subsequently be used to target variable rate application of fertilisers or lime. The soil zones can also be used as a basis for variable seed rate/planting densities, where soil texture within each of the zones is confirmed by sampling and the seed rate/planting density adjusted between the zones based on soil texture differences affecting yields/establishment. For example:

- Areas of different soil textures may have contrasting plant establishment rates and therefore, to create an even crop, the seed rate can be altered to take this into account. Areas of high clay content or high stone content may be expected to have higher plant losses and therefore a higher seed rate is required to achieve an even crop.
- Inherently higher yielding areas may have the potential to support a higher density of plants and therefore justify a higher seed rate.
- Areas of a field which are known to suffer higher pest (i.e. slugs) or weed (i.e. blackgrass in cereal crops) pressure can have a higher seed rate to compensate for this.

Variable seed rate is predominately used in cereal and oilseed crops. Blake et al. (2003) showed that soil texture had a strong influence on percentage establishment in cereal crops, with 60–65% establishment on clays and loams and 90% on sands, indicating that variable rate seeding may be worthwhile in fields with a high degree of topsoil texture variability. However, the methodology may not be directly transferrable to horticultural systems and requires field testing to assess its viability and potential benefits. There are also challenges with adjusting machinery to achieve variable rate seeding/planting (Andrew Richardson, pers. comm.), although it has been attempted in beetroot and vining pea production. Salad growers in Europe have used soil EC maps to allocate small head weight salad crops to lighter areas of a field and larger head varieties to heavier areas.

Limitations of Using Soil Sensor Data to Inform Management

The main limitation of soil mapping using soil sensors for horticultural growers is understanding how the soil maps relate to marketable yields and then how this information can be used to vary soil/crop management. Comparison of soil maps with crop yields is often not possible as most horticultural growers do not collect spatially explicit yield information. However, it may be possible to compare with crop canopy maps (e.g. satellite Normalised Difference Vegetation Index—NDVI—maps as an indicator of crop growth) to look for patterns in variability between soil and crop growth. Where patterns are evident over multiple cropping seasons, this may support a management zone approach.

The relationships between sensor data and the soil property of interest will be highly variable according to soil type, mineralogy and other biophysical factors. The extent to which these relationships have been researched depends on the sensor in question. For example, whereas there has been considerable research effort on defining how measured soil EC relates to soil properties, there is very limited information on the relationship between soil brightness and soil properties. Soil brightness imagery may, therefore, be best used in conjunction with other information (i.e. yield maps/crop satellite imagery or field soil surveys) to define soil management zones, rather than in isolation. Similarly GPR can detect sub-surface features that may be useful for irrigation, such as preferential flow pathways or the depth of shallow groundwater tables, but in very conductive soils (clays or saline soil) the depth penetration of GPR is limited.

Although soil sensors facilitate the estimation of a higher number of soil samples, most sensors still do not measure at all locations and there is a necessity for interpolation. Surveys are clearly a costly operation, so it is important to consider which sampling methods and designs are most cost effective to produce a reliable soil property map; and whether alternative approaches that complement a geostatistical survey may be more cost-effective. Understanding the degree of spatial variability through soil sensing and soil survey may be the most advantageous and cost-effective starting point. Further decision-making should then be guided by the total cost of the survey. For a specific cost, how many samples can be collected and how much time is needed to design the sampling scheme and to collect the field data? These questions are in turn determined by the specific management intervention and the accuracy required to provide relevant data on soil variation.

An alternative approach for measuring spatial variability in soil organic matter, soil pH and nutrients is to use visible-to-near-infrared and mid-infrared 'on-the-go' sensors (e.g. Gholizadeh et al. 2013; Hong et al. 2022). However, this technique requires laboratory calibration for the specific soil types concerned and is not a substitute for a soil type map.

2.6 Current Levels of Adoption

The use of sensors to determine soil spatial variability is available in most parts of the world. For example, many farmers and growers have used grid sampling and soil analysis to determine management zones (based on interpolated data) for the variable rate application of lime, phosphate and potash. Such soil pH and nutrient mapping has been available as a commercial service for many years in Europe. The use of EMI in combination with soil survey is also available as a commercial service, with farmers able to establish management zones based on the physical and drainage properties of the topsoil and subsoil. EMI is also increasingly used in combination with yield mapping to delineate management zones for precision farming (Rodrigues et al. 2015; Karp et al. 2023). Other emerging technologies such as the use of XRF spectroscopy are mainly at the research stage with very limited commercial application in agriculture or horticulture, mainly due to cost. In a review of public data on farm level use of precision agriculture (PA) in crop production worldwide, Lowenberg-deBoer and Erickson (2019) found that there were clear differences in adoption between PA practices. For example, GPS guidance and interdependent technologies have been adopted at the same rate as other historic agricultural technologies, whereas variable-rate technology adoption rarely exceeds 20% of farms although adoption rates can be as high as 50% for certain niche groups of farmers, particularly those with higher soil variability (Schimmelpfennig and Lowenberg-DeBoer 2020). The largest gap in PA adoption is associated with small- and medium-size farms within the developing world that make little use of mechanised farm equipment (e.g. Munz and Schuele 2022). Findings based on corn farms in the US support the fact that economies of scale play a role in the adoption of PA practices. Larger farms showed higher frequency of adoption of GPS based yield and soil mapping, guidance systems and variable-rate application technologies (Schimmelpfennig 2016).

Carrer et al. (2022) corroborate this frequency of adoption by stating that farm size, economies of scale, farmer education and the availability of technical assistance were the main factors influencing the adoption of PA technologies by sugarcane farmers in Brazil. This was supported by Schimmelpfennig (2016) who found that larger corn farms in the U.S. showed higher frequency of adoption of GPS based

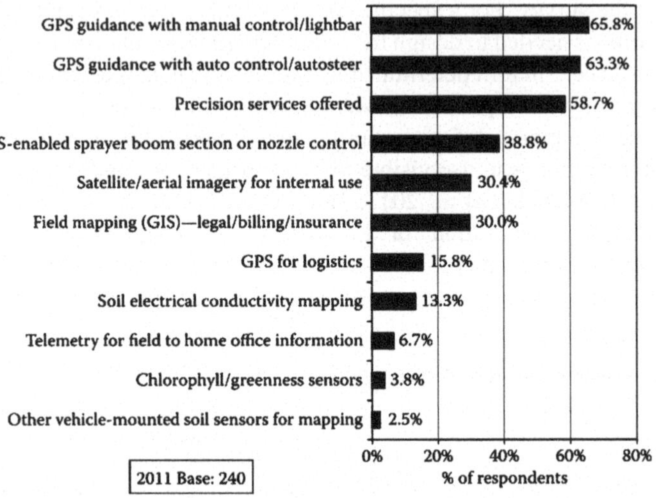

FIGURE 1.8 Summary of a nationwide survey reflecting the use of precision technologies across agricultural states in the United States. (Modified and adapted from Whipker, L. D., and B. Erickson, 2011 Precision agriculture services dealership survey results. Staff paper, Department of Agricultural Economics, Purdue University. W. Lafayette, IN, 2013.)

Fig. 2.5 Summary of a nationwide survey reflecting the use of precision technologies on all farm types across agricultural states in the United States

yield and soil mapping. In India, Narwane et al. (2022) found that significant barriers to PA technology implementation included technological challenges such as technical complexity, but also social challenges such as acceptance by society and user, technical skills, as well as organisational and economic challenges including trust, initial investment and return on investment. Precision agriculture technologies are continuously evolving and therefore changing the significance of these barriers to adoption. For example, Mulla and Khosla (2015) focused on the use of informatics and decision support systems to determine management zones or variable-rate recommendations. They found that the amount and number of data sources that inform PA was increasing, whereas, the scale at which PA management is implemented was decreasing (Fig. 2.5).

Applicability Across Horticultural Cropping Systems

Soil sensing, sampling and surveying can be used to identify soil and land suitability for horticultural cropping, and subsequently to aid management of perennial and annual crops through zoning. Soil suitability for different crops is determined by assessing soil depth, texture, stoniness and drainage characteristics. Soil chemical characteristics can also be important in the case of naturally alkaline soils for crops such as mango (*Mangifera indica*) and pineapple (*Ananas comosus*), and depth to a strongly calcareous layer (high pH) for top fruit such as orange (*Citrus cinensis*) (Landon 2014). For perennial crops, the establishment of soil management zones can therefore be used to determine planting and landscaping schemes, matching crops, margins and soil protection areas (e.g. grassed waterways) according to soil and land suitability.

EMI and other geophysical techniques that essentially identify changes in soil texture (vis moisture) only need to be applied once (as long as they are carried out at an optimal time of the year in the first place). For annual crops, the derived soil management zones can be used as the basis for variable rate application of lime, nutrients and water. Variable rate application of lime, phosphate and potash relies on regular soil sampling and analysis and is relevant for all commercial crops in fields where soil acidity and soil nutrient reserves vary. However, variable rate application of nitrogen is mainly applicable to crops such as cabbages and cauliflowers in which nitrogen applications are split through the season. For such crops, canopy sensing during earlier growth stages can be used to guide subsequent variable nitrogen application rates. Soil information and the establishment of soil management zones to determine spatial variation in physical limitations to crop growth can also be used in combination with canopy sensing to apply variable rates of nitrogen. This combination of techniques can account for variations in soil nitrogen supply and yield potential. However, benefits are only likely to be seen where nitrogen is the dominant growth or quality limiting factor. Other yield and quality factors include other nutrient deficiencies, poor drainage, drought, soil compaction, disease pressure and weed pressure. The commercial value of this approach can therefore be difficult to demonstrate (see Chap. 7 on Crop nutrition).

Barriers to Further Adoption

Factors that inhibit the further adoption of soil sensing, sampling and survey technologies include cost, technical know-how and technical support. More specifically, costs of hardware/sensors (depending on the scale of operation) and access to credit can be a significant barrier, along with data management (processing of big-data and sharing of knowledge, such that it can be interpreted by practitioners); literacy rates of farmers and their willingness to adopt new technologies; connectivity to high-speed internet; hindrance due to a lack of interoperability between digital infrastructures and restricted decision-making processes due to land tenure (Shafi et al. 2019; Kanter et al. 2019). Over 40% of 504 Brazilian farmers surveyed reported that connectivity, lack of investment capital and the literacy rate posed challenges to adoption (Bolfe et al. 2020). Depending on the context, some barriers to uptake are of a more structural nature or subject to change due to socio-economic changes such as market dynamics. For example, Mulla and Khosla (2015) reported that based on the period of 2000–2014, US commodity prices were a major factor that determined PA uptake. That is, a consistent increase in higher economic gains reduced the costs barrier as profits were re-invested into novel technologies and techniques.

To give growers flexibility of management, it is also important that technical support is not tied to the use of specific products. Smaller scale horticulture producers can implement many aspects of precision management through their knowledge of their plot of land. However, at larger scales, soil sensing and sampling technologies become more cost-effective and relevant, particularly on 'new' land that is unfamiliar to a grower. Nevertheless, several barriers to adoption still exist.

Most research on barriers to adoption have focused on technical challenges, while studies on social, economic and organisational factors have been minimal. Narwane et al. (2022) identified technical factors such as technical complexity, energy efficiency (challenging in remote locations); technical skill requirement; social factors (including user acceptance), skill issues and trust; and economic factors (including investment cost) and a lack of overall economic analysis (i.e. cost-benefit).

Nevertheless, soil sensing technologies developed in the European Union or the United States can provide an opportunity for farmers in other parts of the world. In Rwanda for example, in 2020, the Ministry of Agriculture and Animal Resources (MINAGRI) launched a project (funded by the Bill and Melinda Gates Foundation—BMGF) to update Rwandan soil maps and develop location-specific interventions for farmers using soil testing linked to high-resolution geographic information systems (GIS) (Wetzels 2021). However, such projects must be implemented alongside local training, information provision and continued communication with and between small farmers and growers (Ncube et al. 2018; Wetzels 2021). Technology transfer without local capacity building can result in problematic dependencies (Wetzels 2021). For example, use of a sensor technology that is reliant on technical support from overseas and therefore does not result in investment in the local skills base.

The cost of data is another barrier to adoption. It is often the case that commercial farmers can access certain data and technology while smallholders cannot (Ncube et al. 2018). This cost barrier can also make smaller farmers dependent on big tech players and other global companies to provide data or effectively extract data from their farms and use the information to direct product choices, thereby reducing farmer independence and locking them into a company's value chain (FOEE 2020). A 2019 report by the EU-funded Technical Centre for Agricultural and Rural Cooperation (CTA) also warns of this risk to growers in the global south, stating that the presence of global companies may "*bring increased financial, human and technological resources to the sector, and may be accompanied by major investment in important underlying infrastructure*". However, "*their entry does not replace the need for strong local talent. The capabilities of big tech should complement organisations on the ground that are well positioned to design products that can serve the needs of farmers in their region and business models that will work given local conditions*" (Tsan et al. 2019).

Gonzaga et al. (2019) found that in the midwestern region of Brazil, education, technical assistance and knowledge exchange between growers had a positive effect on technology use but did not necessarily favour systems change through the adoption of multiple technologies. Such changes that give rise to increases in productivity and efficiency were favoured by training courses in technology, partnerships with agri-food companies and making resources available for investment. Similarly, in the Jianghan Plain, China, Yang et al. (2022) found that precision technologies were most effectively promoted through improving farmers' and elders' knowledge of their benefits by two main approaches. Firstly, the use of multiple tools (including social media) to promote technologies, and secondly, enabling the exchange of technical information between farmers (i.e. both technical expert/adviser to grower and peer to peer learning). They also emphasised the importance of providing technical instruction at the farm to understand the challenges of technology implementation experienced by growers. Such an approach can enhance grower confidence in mastering the technology and alleviate concerns regarding technology adoption.

2.7 Next Steps and Likely Future Developments

In recent years, there has been a renewed interest in soil sensing and its application at various spatial scales. This is partly due to recent technological developments in sensing, ongoing advances in information and communication systems and their increasing commercial availability at reasonable cost (Viscarra Rossel and Bouma 2016; Abdillah et al. 2023). Soil sensing can form the basis for further investigation of the physical, chemical and biological properties of the whole soil profile as well as their spatial variability; and could potentially be used to monitor soil organic carbon levels. It could therefore help us gauge the potential of soil to provide food, fibre and other ecosystem services. Agricultural landscapes are increasingly expected to deliver ecosystem services (e.g. C sequestration) and minimise ecosystem disservices (e.g. N leaching) in addition to primary production of food. The role that precision technologies can play within this debate relates to the contrast between the economic and environmental benefit. Whereas economic benefits play out at the field and farm-scale, environmental benefits often play out at the landscape scale. Accurate assessment of soil suitability and land capability at field and landscape scales could help determine which parts of a field or farm/nursery should be devoted to food production and which to providing other services, such as biodiversity, water regulation and climate regulation. The appropriate allocation of land use to non-provisioning services according to soil suitability could increase the multifunctionality of landscapes, and complement, enhance and increase the resilience of food production systems (Redhead et al. 2022; Birnbeck et al. 2025).

Lal (2016) acknowledges this potential but also states that the environmental and economic benefits of precision technologies have not been widely quantified, especially for horticultural crops (Alamo et al. 2012). Reducing pressures on the environment is frequently mentioned as a benefit of precision technologies and practices (Alexandridis et al. 2015; Zanella et al. 2015; Hanson et al. 2022), however, there have been very few studies that have quantified the environmental benefits of precision agriculture (Khanal and Lal 2016). This may be because mapping the full spectrum of soil ecosystem services that could be delivered through PA would require monitoring and thus sampling at a variety of geographic scales (Robinson et al. 2013; Maseyk et al. 2017). More research is needed on how environmental variation affects different sensing techniques and what possibilities there are for fusion sensor data to overcome these effects.

As food systems change to accommodate an increased demand for fruit and vegetables there is likely to be a simultaneous increased demand for information on the suitability of land for horticultural production. This may occur at various spatial scales from national to regional and local (i.e. field scale). Such information will not only be useful to plan cropping rotations but is also essential for soil protection planning and management. The approaches and technologies used to provide soil information at each of the spatial scales will vary depending on cost implications, and the level of accuracy and detail required at each scale. Žížala et al. (2022) produced national high resolution (20 m pixel) surface soil property maps (soil organic carbon—SOC, texture, pH, bulk density, soil depth) for the Czech Republic using digital soil mapping (DSM) methods combined with limited existing soil information. Their predictive mapping approach used a random forest model spatial interpolation method (RFSI; Sekulić et al. 2020), which directly examines the target variable values in the nearest *n* points and their corresponding geographic distances from the predicted point; a mosaic of bare soils from Sentinel-2 satellite data; and terrain attributes (e.g. aspect, elevation and slope). Such DSM methods can provide useful information at detailed spatial scales combined with overall accuracy assessments including potential estimation error and confidence intervals for individual mapping units and points. The method, with or without remote sensing data from bare soils, appears to be very useful for mapping soil surface properties at national and regional scales. It can therefore provide useful information for agricultural, environmental and soil protection policy and planning. However, prediction accuracy was less reliable at local (i.e. field and farm/nursery) scales, indicating that the use of 'local models' or soil sensing combined with more traditional soil survey techniques would provide more accurate and agronomy-relevant information. A further limitation is that the method focuses on soil surface properties and while terrain attributes (such as slope and landscape position, derived from digital terrain models) can provide an indication of soil drainage properties, reliable information on the subsoil is needed to determine key information on yield potential.

One of the key factors determining yield potential is soil water holding capacity as a function of soil depth. The influence of soil hydrology on crop growth and environmental impacts was highlighted by Mulla et al. (1996). They found that the risk of pesticide leaching losses to groundwater could mostly be attributed to variations in soil water movement (i.e. spatial patterns of soil hydraulic properties) rather than varying concentrations of soil organic carbon or other soil properties.

This finding illustrates the potential limitations of inferring soil functions and services from map-derived soil surface characteristics; see Baveye and Laba (2015) for further discussion.

There is therefore great potential to use soil sensing and survey technologies across the world to optimise land use and management within horticultural cropping systems. However, realising this potential is dependent on having appropriate social, economic and technical support to enable uptake. Useful technological approaches can be developed by making funding available through 'eco-schemes'. This is already happening in countries such as the United States and Europe (e.g. through the Common Agricultural Policy) and could be further enhanced (Ncube et al. 2018). However, the sustained use of appropriate technologies in low-income countries could also be supported with targeted funding and could be secured through appropriate technology transfer combined with local capacity building in terms of expertise, training, consultancy and 'peer-to-peer' learning.

Case Study 2.1: Investigating the Value of Uniform- Management Zone and Variable-Rate Based Fertiliser Applications Based on Soil Spectroscopy Data

Headline: This case study considers the associated uncertainty of soil data estimated by spectroscopy and its effect on decision-making in precision horticulture.

Methods*:* In theory, the application of precision agriculture facilitates both a reduction of farm inputs together with an increase in profits. Thus, adoption of targeted management practices should be assessed based on those two outcomes. This case-study investigated the cost-benefit of using management zones or geostatistics to inform variable-rate application of phosphate fertiliser in horticultural fields in Cambridgeshire, UK. We extended the case study of Breure et al. (2022) which expressed the value of PA by means of a loss function framework. We refer the reader to this publication for further details on the methodology. In its generic form, the loss is defined as the difference between the result based on estimated information compared to 'perfect' information given by:

$$L(F) = \Phi(F_0) - \Phi(F)$$

Where $\Phi(F_0)$ is the profit obtained given the fertiliser rate based on 'perfect' information, F_0, which in this case is the soil property values estimated from soil spectra without an error variance. The profit based on estimated soil property values ($\Phi(F)$) can take different values, depending on the sampling approach. In this study, F takes three different values: F_{UA} stands for uniform application based on an estimate of the field mean, F_{opt} stands for the optimised amount of fertiliser to apply, considering the uncertainty around the estimate based on geostatistics, F_{mz} stands for the amount of fertiliser to apply based on a mean estimate for each management zone. Profits are obtained by accounting for the price of fertiliser and the economic value of the crop in published dose-response curves (Prasad et al. 1988). The expected financial loss is then subsequently computed for these different fertiliser regimes and associated uncertainty in soil estimates. See Breure et al. (2022) for further details.

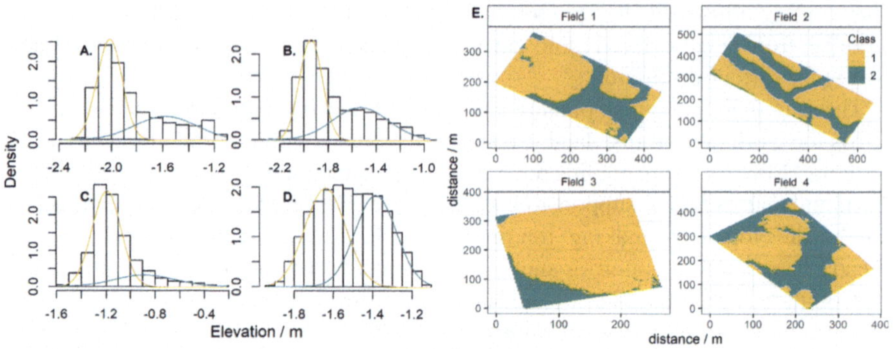

Fig. 2.6 Density distributions of elevation as measured by LiDAR for the whole field together with the normal distributions corresponding to each zone (colours) for fields 1 (**a**) to 4 (**d**). Panel (**e**) shows the distinguished zones as a spatial unit for each field. Note that the fields are located in an area that has been drained, hence the negative elevation values

Whereas the original case-study assessed the difference between uniform- and variable-rate application based on geostatistics, for the purpose of this chapter it has been extended by including a management zone approach. Based on pedological knowledge of the case-study area, elevation as measured by LiDAR (light detection and ranging) is a proxy of two dominant soil types: alluvial/marine silts (drainic sapric Histosol) and peat (mollic Gleysol) (IUSS Working Group WRB 2015). The management zones were defined by applying the expectation-maximisation algorithm on the LiDAR data, which allows for representing its density by two separate normal distributions (Benaglia et al. 2009) (Fig. 2.6).

In this case, soil management zones were treated as a homogenous unit, and it was important that the samples taken to represent these zones captured the within-zone variation. Therefore, balanced sampling was performed on the x and y coordinates (scaled to unit variance) and LiDAR (Grafström and Lisic 2019). The balanced sampling was performed for a range of sample sizes ($n = 4$–30, with intervals of 2), given that the geostatistical approach used estimated soil property data derived from a calibration set sample size where $n = 30$.

Results and Discussion

Findings showed how management zones compared to uniform- and variable-rate application approaches. Management zones were not more cost-effective than geostatistics for fields 1 and 2, and in fact showed in some cases a higher expected loss compared to uniform application (Table 2.1).

These findings are most likely due to the fact that phosphorus (P) concentrations were not well correlated with the LiDAR data for those fields ($\rho = 0.37$ for field 1 and $\rho = 0.01$ for field 2). As a result, the management zones did not reduce

Table 2.1 Adapted table from Breure et al. (2022): fertiliser used (kg ha^{-1}) for perfect knowledge (F_0), variable-rate application (F_{opt}) and uniform application based on wet chemistry samples (F_{UA}) and management-zone approach (F_{mz})

		Fertiliser (kg P_2O_5 ha^{-1})				Expected loss (£ ha^{-1})		
Field	Area (ha)	F_0	F_{opt}	F_{UA}	F_{mz}	$E[L(F_{opt})]$	$E[L(F_{UA})]$	$E[L(F_{mz})]$
1	8.2	113	113	120	116—120	31	38	38—42
2	16.9	102	103	120	112—120	38	52	52—57
3	5.1	103	104	61	98—112	33	154	52—65
4	8.9	47	53	57	30—66	94	145	120—135

Expected loss (relative to perfect knowledge) (£ ha^{-1}) for variable-rate application, $E[L(F_{opt})]$, uniform application, $E[L(F_{UA})]$ and management-zone approach, $E[L(F_{mz})]$. The range of values for $E[L(F_{mz})]$ corresponds to the range obtained from different sample sizes (n = 4–30, with intervals of 2)

variability of P within zones compared to the whole-field variability of P. Neither was the variation in P substantially different between zones for both fields (Fig. 2.7).

Farm management history is a likely driver of these differences, as a series of uniform applications can mask the expected differences in soil nutrients between the Histosols and Gleysols. In this case, a combination of the timing of fertiliser application and the soil's P sorption capacity determines whether a field is suitable for variable-rate application. Furthermore, the case-study also illustrated that the decision on the covariate to inform zones is important. That is, a given covariate should describe the agronomic property well, independent of variations in time (Whelan 2018). In contrast to fields 1 and 2, the expected loss from a management zone approach lay in between the geostatistical- and uniform application approach for fields 3 and 4 (Table 2.1). This is primarily due to the fact that management zones successfully reduced variability of P compared to the whole field variation and were substantially different from each other (Fig. 2.7). Therefore, the covariate (LiDAR) could reduce variability for an agronomic property of interest (soil P) that was not used to classify the field into management zones. We ascribe this to a good correlation between the LiDAR and available soil P ($\rho = 0.87$ for field 3 and $\rho = -0.73$ for field 4). Lastly, we note that there are substantial differences in the quantity of fertiliser applied between the different approaches. Of particular importance is that increased profitability from variable-rate application does not strictly coincide with a reduction of fertiliser compared to a uniform application.

Whereas in this case-study the differences between sampling methods were assessed through the associated profit for soil property estimates, Muhammed et al. (2017) compared uniform, management zones and geostatistical approaches based on yield data. They found that on average across the fields studied, the geostatistical approach led to higher profits compared to the management zone approach. The spatial structure of yield maps proved to be a good indicator of profitability. Fields in which the variogram had a large effective range (spatially correlated over longer distances) or high sill (large within-field variance) had a higher probability of being profitable if managed by a geostatistical approach. It is important to note that the data used to delineate management zones in this study was

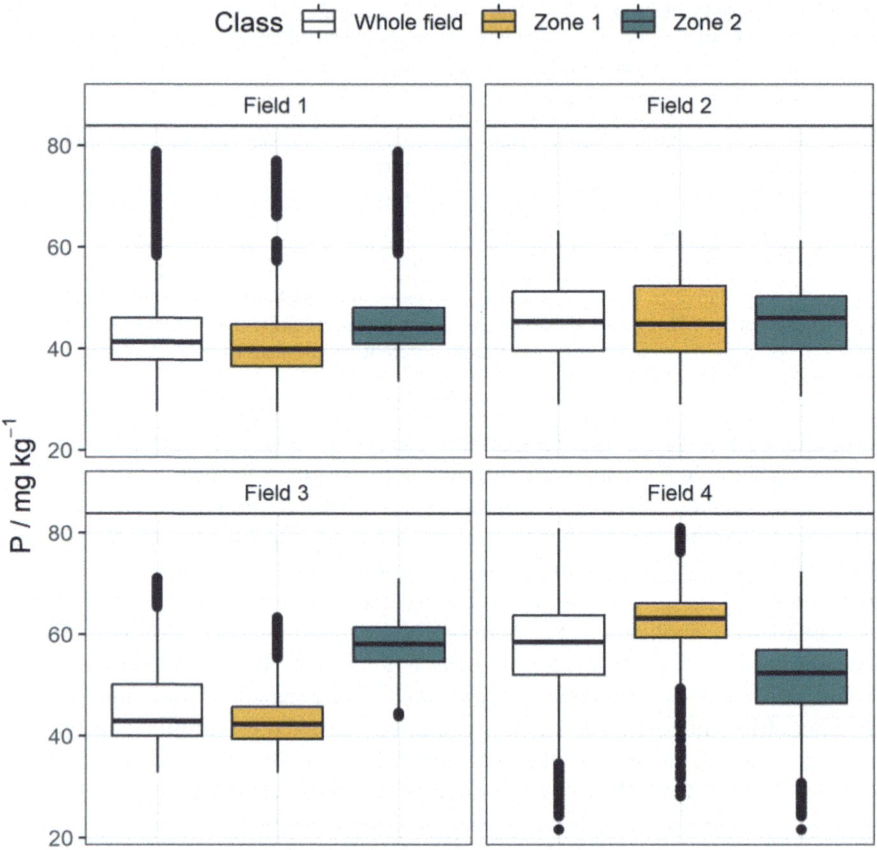

Fig. 2.7 Boxplots of available soil P per management zone and the whole field

LiDAR data and in Muhammed et al. 2017 it was yield data. Delineated zones based on soil nutrient levels might increase the cost-effectiveness of a management zone approach. Lastly, Muhammed et al. (2017) took into account the laboratory costs of soil analysis whereas this case-study disregarded these but accounted for associated uncertainty. We note that for a robust estimate of cost-effectiveness, the costs of both sampling and laboratory costs should be taken into consideration.

Case Study 2.2: Mapping Soil Variability at F.B. Parrish & Son in Bedfordshire, UK

Headline: The soil mapping case study at F.B. Parrish & Son in Bedfordshire highlights the options available to growers wanting to better understand and map their soil variability.

Background: F. B. Parrish and Son, based in Bedfordshire, southeast England in the UK, grow around 480 ha of cereals, 160 ha of alliums (onions and shallots), 60 ha of potatoes and 60 ha of Quinoa in a rotation of onions/shallots, wheat, potato, wheat and Quinoa.

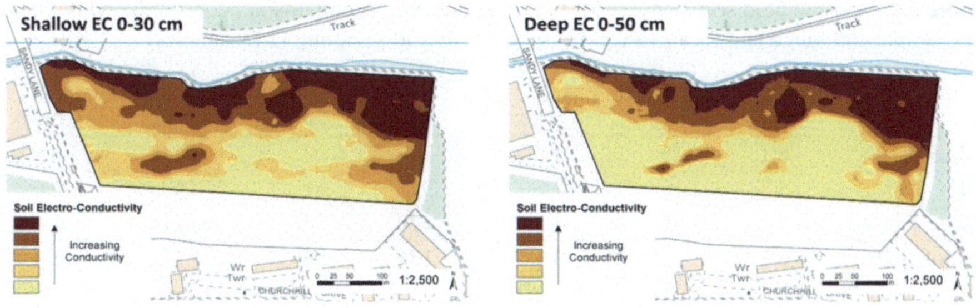

Fig. 2.8 Soil electrical conductivity (EC) maps of avenue field

Avenue field is a 10-ha field on the farm in Bedfordshire, southeast England. The field was used as a case study to demonstrate the options available to growers for soil mapping, including soil EC mapping and soil nutrient mapping. Soil samples were taken from the field to enable comparison of the effect of soil sampling intensity and soil sampling method (a regular grid-based approach compared with a zone-based approach) on the soil pH and nutrient maps produced.

The soils in the area have developed from glacial and river outwash deposits resulting in high variability in soil types from deep clays to deep loamy sands. Recognising the high within-field variability in soils, the farm has had the electrical conductivity (EC) of all fields mapped. The EC maps were used in combination with the existing knowledge on soil and crop variability to define soil 'zones' within each field. These zones are sampled separately for soil pH, P, K and Mg every 4 years. The farm uses the soil pH and nutrient maps for variable rate lime and fertiliser applications. The soil EC maps are also used as the basis for variable rate drilling of the farms arable crops.

EC mapping: Information on soil EC was collected from Avenue field at two soil depths: 0–30 cm 'shallow' and 0–50 cm 'deep'. The EC scanner was driven across the field at bout widths of 12 m and the EC measurements were used to produce soil EC maps (Fig. 2.8).

The soil EC survey identified lighter textured areas on the south side of the field and heavier textured areas on the north and north-west side of the field, corresponding well with the farmers existing understanding of soil texture variability within this field. Soil EC maps can be used to identify different soil management zones for targeted soil sampling or as the basis for variable rate management between light and heavier textured areas within a field (i.e. variable rate planting).

2.8 Soil Nutrient Mapping

Nutrient management guidance for horticultural cropping in England and Wales recommends soil sampling for pH, P, K and Mg every 3–5 years. Traditional soil sampling, where a single composite soil sample is taken from a field (usually a

Table 2.2 Avenue field soil analysis for pH, P, K and Mg—comparison between the whole field soil sample and range and mean values from intensive grid sampling (143 samples)

	pH	P		K		Mg	
		mg/l	Index	mg/l	Index	mg/l	Index
Mean	6.1	35	3	217	3	110	3
Min	5.3	16	2	92	1	53	2
Max	7.1	55	4	428	4	215	4
Whole field	6.1	33	3	171	2-	77	2

composite sample taken from a 'W' across the field) will provide average soil pH, and nutrient indices for the field. This is the cheapest soil sampling option; however, it can potentially conceal variation in soil pH and nutrient indices.

To illustrate the different approaches to soil sampling and mapping, soil samples were taken from Avenue field as follows:

- Single composite sample (of 25 cores) taken walking a 'W' across the field.
- 1 ha soil sampling—a single composite sample (of 25 soil cores) was taken walking a 'W' from each 1 ha block in the field (10 samples in total).
- Grid soil sampling—topsoil samples were taken on a 25 m grid (single composite sample taken from each grid point consisting of 16 soil cores taken in a spiral within a 3 m radius of the central point—143 samples in total).

Soil was analysed for pH (in water), Olsen extractable P, and ammonium nitrate extractable K and Mg. The single 'whole field' sample provided a good measure of the mean field value for pH and P Index but underestimated mean soil K and Mg indices. The more intensive grid soil sampling showed within-field variation in soil pH of 5.3–7.1; soil P Index 2–4 (16–55 mg P/l); and soil K Index 1–4 (92–428 mg K/l; Table 2.2). Analysis of these samples was used to 'map' soil pH and nutrient variability in Avenue field using a number of different approaches and sampling intensities, discussed below.

2.9 Approaches to Regular or 'Grid' Soil Sampling

A regular sampling approach typically divides the field into approximately equal sized areas or blocks and then either the centre point of each area is sampled ('point' sampling) or a representative sample from the whole area is taken (by sampling in a 'W' pattern across the area). Where a representative sample is taken from each area, the soil nutrient map produced will show the boundaries of each sampling area (Fig. 2.9f).

Where a sample is taken from the central 'point' of each area, each sample will be a composite of typically 16 cores taken in a small radius around the pre-designated point. The location of each sampling point is GPS logged and this information is used to create a contoured map of soil pH and nutrients using interpolation to

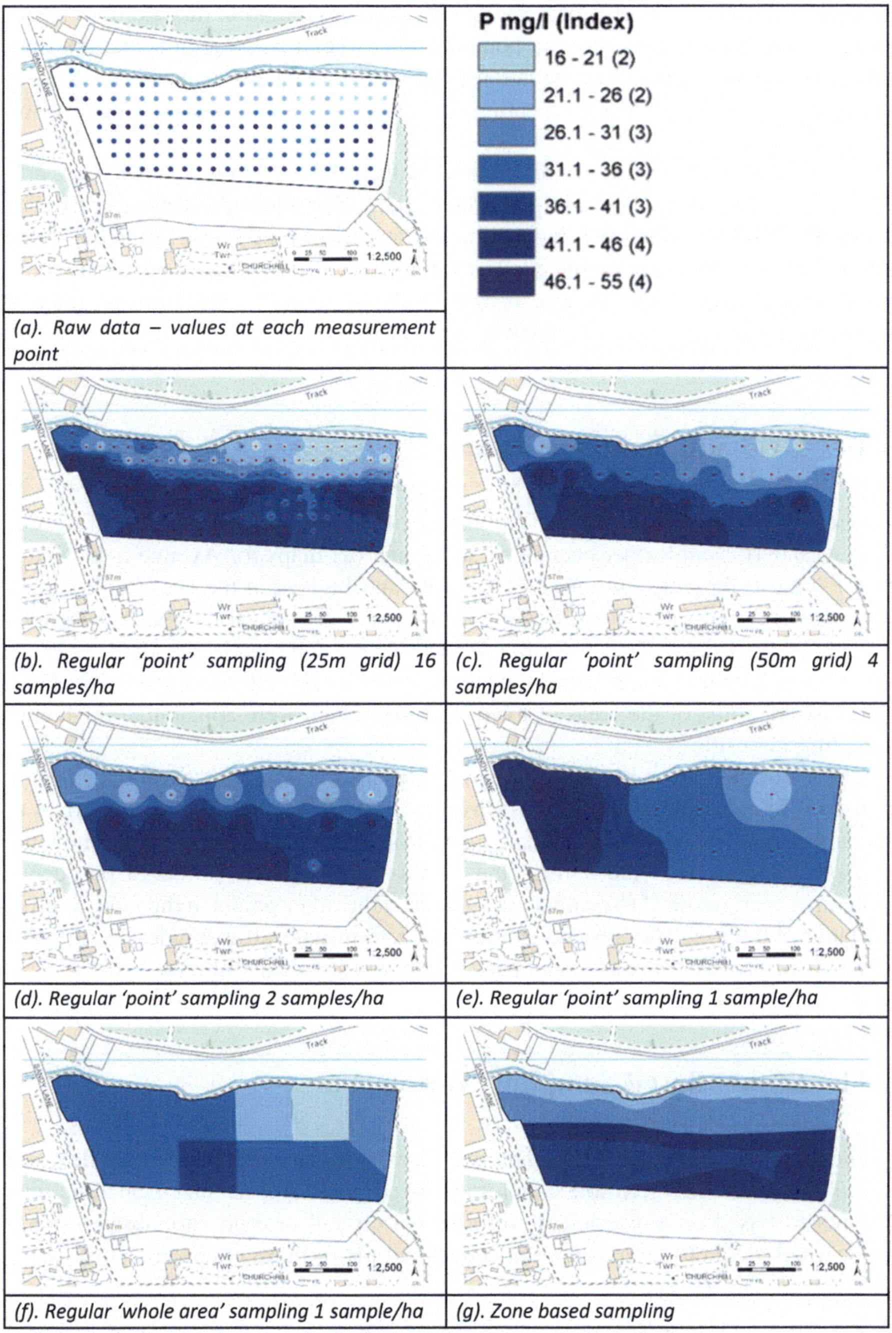

Fig. 2.9 Mapping soil P at Avenue field. (**a**). Raw data—values at each measurement point. (**b**). Regular 'point' sampling (25 m grid) 16 samples/ha. (**c**). Regular 'point' sampling (50 m grid) 4 samples/ha. (**d**). Regular 'point' sampling 2 samples/ha. (**e**). Regular 'point' sampling 1 sample/ha. (**f**). Regular 'whole area' sampling 1 sample/ha. (**g**). Zone based sampling

estimate values between the measured points. The interpolation method uses values at each sample point to determine boundaries between mapping units and does not take any account of soil type variations, but GPS referencing makes it possible to return to the same point when repeating the soil sampling to monitor changes in soil pH/nutrient content over time.

In the UK, the most common commercially used sampling intensity for grid 'point' sampling is one sample per hectare (i.e. sampling on a 100 m grid). The number of samples taken will have an important effect on the level of detail in the soil pH or nutrient maps produced. The limiting factor is normally cost—the more samples that are taken the more detailed the soil map is likely to be, however the cost of additional samples may not always be justified.

2.10 Grid Soil Sampling Results

Figures 2.9b–f and 2.10b–f show the impact of grid 'point' sampling intensities of 1, 2, 4, and 16 samples per hectare on soil P and pH maps for Avenue field. Soil P was lowest in the north/northeast of the field and highest in the south of the field, reflecting soil type differences. Soil pH showed high spatial variability, with areas of low pH on both sides of the field. In this instance the commercial standard sampling intensity of 1 sample/ha identified the lower area of soil P in the northeast of the field, but 'missed' some of the areas of lower soil pH, which were apparent at the higher sampling intensities.

In each of these examples the contoured map of soil P and pH was produced using a method of data interpolation called Inverse Distance Weighting. It is important to remember that where soil samples are taken from a specific point, we only know the soil pH and nutrient values from the sample taken at that point; to produce a contoured map we need to estimate the values at the other points in the field. Inverse Distance Weighting, Nearest neighbour and Kriging are all examples of different methods of data interpolation, and each will produce a different map.

2.11 'Zone' Based Sampling Approaches

'Zone' based, or 'targeted' soil sampling uses existing knowledge of within field soil variability to direct where soil samples are taken. Field soil management zones can be created based on any available information on soil or crop variability within a field, which is likely to impact on or reflect soil pH or nutrient content, such as:

- Soil EC maps.
- Satellite soil brightness maps, which provide a measure of the combined effects of soil texture, organic matter content and soil moisture at the time the image was taken.

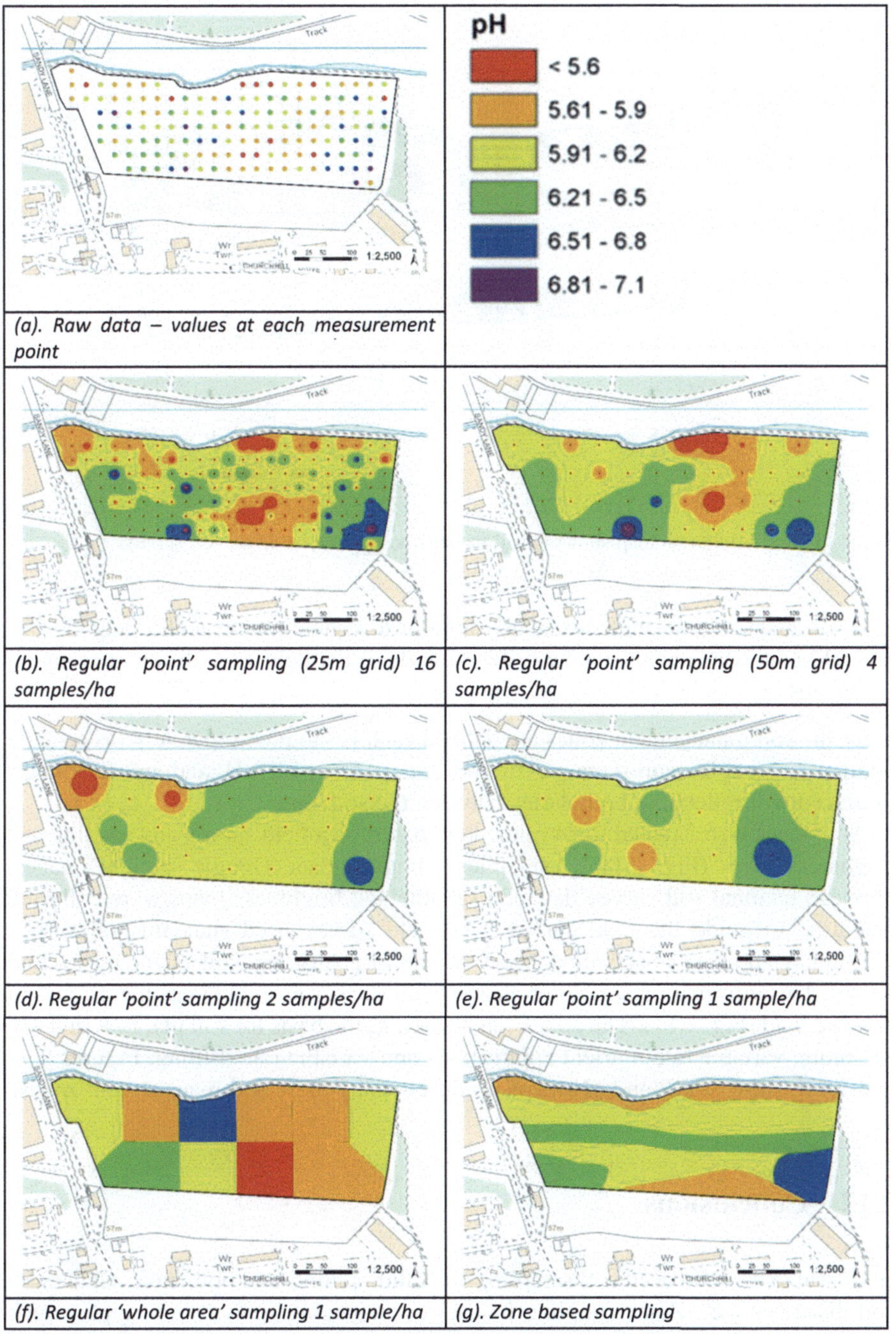

Fig. 2.10 Mapping soil pH at Avenue field. (**a**). Raw data—values at each measurement point. (**b**). Regular 'point' sampling (25 m grid) 16 samples/ha. (**c**). Regular 'point' sampling (50 m grid) 4 samples/ha. (**d**). Regular 'point' sampling 2 samples/ha. (**e**). Regular 'point' sampling 1 sample/ha. (**f**). Regular 'whole area' sampling 1 sample/ha. (**g**). Zone based sampling

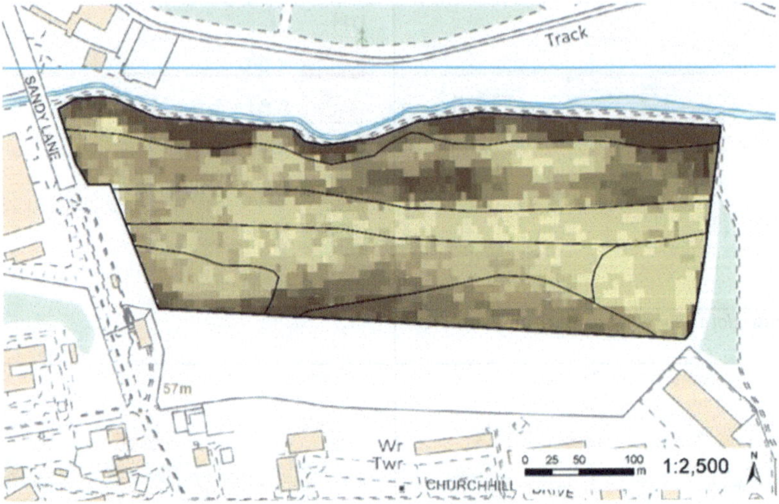

Fig. 2.11 Soil brightness map showing soil management zones based on soil survey

- Yield maps identifying consistently high or low yielding areas of a field
- Field boundary maps—where different parts of the field have different cropping histories.

Once the soil management zones are defined each is sampled separately (as a single composite sample representative of the zone). The soil pH and nutrient maps produced will reflect the boundaries between the soil zones.

Soil zones were created for Avenue field based on a field survey and satellite soil brightness maps (Fig. 2.11). Available geological data, Google satellite images, previous national soil survey data and satellite soil brightness maps were collected and used to guide the field survey. The field survey used standard soil survey techniques and one observation point per hectare (using a soil auger up to 1 m depth). This information was combined to delineate seven soil management zones in Avenue field (Fig. 2.11). The zones can be used as the basis for soil pH, P, K and Mg sampling, variable rate nutrient applications and variable rate drilling. Figures 2.9g and 2.10g show soil P and pH maps produced using these soil management zones.

2.12 Conclusions

Both grid and zone-based soil sampling are valid options, and both have advantages and disadvantages. Zone sampling focuses on managing areas by soil type, which forms the basis of good nutrient management. It uses patterns and boundaries evident from looking at soil surveys or yield maps to form the basis of management zones. However, grid sampling may identify 'hot spots' of soil fertility or pH (often related to field management history) that cannot be achieved using zone sampling; and grid sampling (because of the smaller point area sampled) is better at detecting change over time. Grid sampling is typically more expensive than zone sampling as a greater number of samples is usually taken, and in fields with significant soil variability the number of samples required to create an 'accurate' map may be uneconomic.

Competing Interests The authors have no conflicts of interest to declare that are relevant to the content of this chapter.

References

Abdillah IT, Kurniawati A, Rumanti AA (2023) Factors influencing the adoption of precision agriculture technology (PAT) for organization: a preliminary study. In: Proceedings of the 2023 5th international conference on management science and industrial engineering

Adamchuk VI, Viscarra Rossel RA (2010) Development of on-the-go proximal soil sensor systems. In: Viscarra Rossel RA, McBratney AB, Minasny B (eds) Proximal soil sensing. Springer, New York, pp 15–28

Alamo S, Ramos MI, Feito FR, Cans JA (2012) Precision techniques for improving the management of olive groves of southern Spain. Span J Agric Res 10(3):583–595

Alexandridis TK, Galanis G, Kalopesa E, Cherif I, Chouzouri A, Dimitrakos A, Kalopesas C, Navrozidis I, Patakas A, Thomidis T, Zalidis G (2015) Promoting precision farming in South-east Europe: examples from site-specific management clusters in North Greece. Precis Agric 15: 733–740

Allbed A, Kumar L (2013) Soil salinity mapping and monitoring in arid and semi-arid regions using remote sensing technology: a review. Adv Remote Sens 2013(2):373–385

Arrouays D, Marchant BP, Saby NPA, Meersmans J, Orton TG, Martin MP, Bellamy PH, Lark RM, Kibblewhite M (2012) Generic issues on broad-scale soil monitoring schemes: a review. Pedosphere 22:456–469

Baveye PC, Laba M (2015) Moving away from the geostatistical lamppost: why, where, and how does the spatial heterogeneity of soils matter? Ecol Model 298:24–38. https://doi.org/10.1016/j.ecolmodel.2014.03.018

Benaglia T, Chauveau D, Hunter DR (2009) An EM-like algorithm for semi- and nonparametric estimation in multivariate mixtures. J of Comp and Graph Stat 18:505–526

Benedet L, Faria WM, Silva SHG, Mancini M, Demattê JAM, Guilherme LRG, Curi N (2020) Soil texture prediction using portable X-ray fluorescence spectrometry and visible near-infrared diffuse reflectance spectroscopy. Geoderma 376:114553

Besson A, Cousin I, Samouëlian A, Boizard H, Richard G (2004) Structural heterogeneity of the soil tilled layer as characterized by 2D electrical resistivity surveying. Soil Tillage Res 79:239–249

Birnbeck S, Burmeister J, Wolfrum S, Panassiti B, Walter R (2025) Riparian buffer strips promote biomass, species richness and abundance of flying insects in agricultural landscapes. Agric Ecosyst Environ 378:109300. https://doi.org/10.1016/j.agee.2024.109300

Blake J, Spink JH, Dyer C (2003) Factors affecting cereal establishment and its prediction. AHDB Cereals and Oilseeds

Bolfe ÉL et al (2020) Precision and digital agriculture: adoption of technologies and perception of Brazilian farmers. Agriculture 10:653. https://doi.org/10.3390/agriculture10120653

Boll J, Van Rijn RPG, Weiler KW, Ewen JA, Daliparthy J, Herbert SJ, Steenhuis TS (1996) Using ground-penetrating radar to detect layers in a sandy field soil. Geoderma 70:117–132

Breure TS, Haefele SM, Hannam JA, Corstanje R, Webster R, Moreno-Rojas S, Milne AE (2022) A loss function to evaluate agricultural decision-making under uncertainty: a case study of soil spectroscopy. Prec Agric:1–21

Carrer MJ, de Souza Filho HM, Vinholis MDMB, Mozambani CI (2022) Precision agriculture adoption and technical efficiency: an analysis of sugarcane farms in Brazil. Technol Forecast Soc Chang 177:121510

Clairotte M, Grinand C, Kouakoua E, Thébault A, Saby NPA, Bernoux M, Barthès BG (2016) National calibration of soil organic carbon concentration using diffuse infrared reflectance spectroscopy. Geoderma 276:41–52

Du C, Wang J, Zhou Z, Shen Y, Zhou J (2015) In situ measurement of ammonia concentration in soil headspace using Fourier transform mid-infrared photoacoustic spectroscopy. Pedosphere 25(4):605–612

Escadafal R (1989) Remote sensing of arid soil surface color with Landsat thematic mapper. Adv Space Res 9(1):159–163

Friends of the Earth Europe (2020) Digital Farming. Can digital farming really address the systemic causes of agriculture's impact on the environment and society, or will it entrench them? https://friendsoftheearth.eu/wp-content/uploads/2020/02/foee-digital-farming-paperfeb-2020.pdf

Galambošová J, Macák M, Rataj V, Barát M, Misiewicz PA (2020) Determining trafficked areas using soil electrical conductivity – a pilot study. Acta Technol Agric 23:1–6

Gholizadeh A, Borůvka L, Saberioon M, Vašát R (2013) Visible, near-infrared, and mid-infrared spectroscopy applications for soil assessment with emphasis on soil organic matter content and quality: state-of-the-art and key issues. Appl Spectrosc 67:1349–1362

Gholizadeh A, Žižala D, Saberioon M, Borůvka L (2018) Soil organic carbon and texture retrieving and mapping using proximal, airborne and Sentinel-2 spectral imaging. Remote Sens Environ 218:89–103

Gonzaga JF, Vilpoux OF, Gomes Pereira MW (2019) Factors influencing technological practices in the Brazilian agrarian reform. Land Use Policy 80:150–162. https://doi.org/10.1016/j.landusepol.2018.10.005

Grafström A, Lisic J (2019) Balanced sampling: balanced and spatially balanced sampling. R Package Vers 1(5):5

Grisso R, Alley M, Holshouser D, Thomason W (2009) Precision farming tools: Soil electrical conductivity. Virginia Cooperative Extension Publication 442–508. Available at: https://www.pubs.ext.vt.edu/content/dam/pubs_ext_vt_edu/442/442-508/442-508_pdf.pdf

Hanson ED, Cossette MK, Roberts DC (2022) The adoption and usage of precision agriculture technologies in North Dakota. Technol Soc 71:102087. https://doi.org/10.1016/j.techsoc.2022.102087

Heege H (2013) Sensing of natural soil properties. In: Heege H (ed) Precision in crop farming: site specific concepts and sensing methods: applications and results. Springer, pp 51–102

Hong Y, Munnaf MA, Guerrero A, Chen S, Liu Y, Shi Z, Mouazen AM (2022) Fusion of visible-to-near-infrared and mid-infrared spectroscopy to estimate soil organic carbon. Soil Tillage Res 217:105284. https://doi.org/10.1016/j.still.2021.105284

Huisman JA, Hubbard SS, Redman JD, Annan AP (2003) Measuring soil water content with ground penetrating radar: a review. Vadose Zone J 2(4):476–491. https://doi.org/10.2113/2.4.476

IUSS Working Group WRB (2015) World Reference Base for Soil Resources 2014, update 2015 International soil classification system for naming soils and creating legends for soil maps. World Soil Resources Reports No. 106. FAO, Rome

Jenny H (1941) Factors of soil formation: a system of quantitative pedology. Dover Publications, New York, p 281

Johnson JM, Vandamme E, Senthilkumar K, Sila A, Shepherd KD, Saito K (2019) Near-infrared, mid-infrared or combined diffuse reflectance spectroscopy for assessing soil fertility in rice fields in sub-Saharan Africa. Geoderma 354:113840

Kalnicky DJ, Singhvi R (2001) Field portable XRF analysis of environmental samples. J Hazard Mater 83:93–122. https://doi.org/10.1016/S0304-3894(00)00330-7

Kanter DR, Bell AR, Shukla McDermid S (2019) Precision agriculture for smallholder nitrogen management. One Earth 1:281–284

Karp FH, Adamchuk VI, Melnitchouck A, Allred B, Dutilleul P, Martinez LR (2023) Validation and potential improvement of soil survey maps using proximal soil sensing. J Environ Eng Geophys

Khanal S, Lal R (2016) Precision agriculture for improving water quality under changing climate. In: Lal RA, Stewart BA (eds) Soil-specific farming precision agriculture. CRC Press, Boca Raton, pp 283–306

King JA, Dampney PMR, Lark RM, Wheeler HC, Bradley RI, Mayr T, Russill N (2003) Evaluation of non-intrusive sensors for measuring soil physical properties. Project Report 302, HGCA, London

Knox NM, Grunwald S, McDowell ML, Bruland GL, Myers DB, Harris WG (2015) Modelling soil carbon fractions with visible near-infrared (VNIR) and mid-infrared (MIR) spectroscopy. Geoderma 239:229–239

Krajčo J (2007) Detection of soil compaction using soil electrical conductivity. PhD thesis, Cranfield University

Kuang B, Mahmood HS, Quraishi MZ, Hoogmoed WB, Mouazen AM, van Henten EJ (2012) Sensing soil properties in the laboratory, in situ and on-line. A review. Adv Agric 114:155–223

Kweon G (2012) Toward the ultimate soil survey: sensing multiple soil and landscape properties in one pass. Agron J 104:1547–1557

Lal R (2016) Challenges and opportunities in precision agriculture. In: Lal RA, Stewart BA (eds) Soil-specific farming precision agriculture. CRC Press, Boca Raton, pp 391–400

Lambot S, Slob E, Rhebergen J, Lopera O, Zaib Jadoon K, Vereecken H (2009) Remote estimation of the hydraulic properties of a sand using full-waveform integrated Hydrogeophysical inversion of time-lapse, off -ground GPR data. Vadose Zone J 8:743–754. https://doi.org/10.2136/vzj2008.0058

Landon JR (2014) Booker tropical soil manual: a handbook for soil survey and agricultural land evaluation in the tropics and subtropics, vol 530. Routledge, London. https://doi.org/10.4324/9781315846842

Lark RM (2000) Estimating variograms of soil properties by the method-of-moments and maximum likelihood. Eur J Soil Sci 51:717–728

Lark RM, Cullis BR, Welham SJ (2006) On spatial prediction of soil properties in the presence of a spatial trend: the empirical best linear unbiased predictor (E-BLUP) with REML. Eur J Soil Sci 57:787–799

Liu X, Dong X, Leskovar DI (2016) Ground penetrating radar for underground sensing in agriculture: a review. Int Agroph 30:533–543

Lombardi F, Ortuani B, Facchi A, Lualdi M (2022) Assessing the perspectives of ground penetrating radar for precision farming. Remote Sens 14(23):6066. https://doi.org/10.3390/rs14236066

Lowenberg-DeBoer J, Erickson B (2019) Setting the record straight on precision agriculture adoption. Agron J 111(4):1552–1569

Lund ED, Christy CD, Drummond PE (2000) Using yield and soil electrical conductivity (EC) maps to derive crop production performance information. In: Proceedings of the 5th international conference on precision agriculture

Marchant BP, Newman S, Corstanje R, Reddy KR, Osborne TZ, Lark RM (2009) Spatial monitoring of a non-stationary soil property: phosphorus in a Florida water conservation area. Eur J Soil Sci 60:757–769

Marchant BP, Saby NPA, Lark RM, Bellamy PH, Jolivet CC, Arrouays D (2010) Robust analysis of soil properties at the national scale: cadmium content of French soils. Eur J Soil Sci 61:144–152

Marques da Silva JR, Rebollo FJ, Sousa A, Mesquita P (2012) Yield potential probability maps using the Rasch model. Biosyst Eng 111(4):369–380. https://doi.org/10.1016/j.biosystemseng.2012.01.002

Maseyk FJ, Mackay AD, Possingham HP, Dominati EJ, Buckley YM (2017) Managing natural capital stocks for the provision of ecosystem services. Conserv Lett 10(2):211–220

Monaghan J (2015) Towards precision inputs and improved understanding of the underlying causes of in-field variation in Lettuce crop maturity and yield. Annual Report for AHDB Horticulture project CP 121

Muhammed SE, Marchant BP, Webster R, Whitmore AP, Dailey G, Milne AE (2017) Assessing sampling design for determining fertilizer practice from yield data. Comp Electr Agric 135:163–174

Mulla D, Khosla R (2015) Historical evolution and recent advances in precision farming. In: Lal RA, Stewart BA (eds) Soil-specific farming precision agriculture. CRC Press, Boca Raton, pp 1–35

Mulla DJ, Perillo CA, Cogger CG (1996) A site-specific farm-scale GIS approach for reducing groundwater contamination by pesticides. J Environ Qual 25(3):419–425

Munz J, Schuele H (2022) Influencing the success of precision farming technology adoption—a model-based investigation of economic success factors in small-scale agriculture. Agriculture 12(11):1773. https://doi.org/10.3390/agriculture12111773

Narwane VS, Gunasekaran A, Gardas BB (2022) Unlocking adoption challenges of IoT in Indian agricultural and food supply chain. Smart Agric Technol 2:100035

Ncube B, Mupangwa W, French A (2018) Precision agriculture and food security in Africa. In: Mensah P, Katerere D, Hachigonta S, Roodt A (eds) Systems analysis approach for complex global challenges. Springer, Cham

Ng W, Minasny B, Montazerolghaem M, Padarian J, Ferguson R, Bailey S, McBratney AB (2019) Convolutional neural network for simultaneous prediction of several soil properties using visible/near-infrared, mid-infrared, and their combined spectra. Geoderma 352:251–267

Oldoni H, Rodrigues Tavares T, Luis Brasco T, Roberto Cherubin M, Pereira de Carvalho HW, Graziano Magalhães PS, Rios do Amaral, L. (2024) Temporal evaluation of soil chemical quality using VNIR and XRF spectroscopies. Soil Tillage Res 240:106087. https://doi.org/10.1016/j.still.2024.106087

Oliver MA, Webster R (2014) A tutorial guide to geostatistics: computing and modelling variograms and kriging. Catena 113:56–69

O'Rourke SM, Minasny B, Holden NM, McBratney AB (2016a) Synergistic use of Vis-NIR, MIR and XRF spectroscopy for the determination of soil geochemistry. Soil Sc Soc Am J 80:888–899

O'Rourke SM, Stockmann U, Holden NM, McBratney AB, Minasny B (2016b) An assessment of model averaging to improve predictive power of portable vis-NIR and XRF for the determination of agronomic soil properties. Geoderma 279:31–44

Prasad M, Spiers TM, Ravenwood IC (1988) Target phosphorus soil test values for vegetables. New Zeal J Exp Agric 16:83–90

Redhead JW, Hinsley SA, Botham MS, Broughton RK, Freeman SN, Bellamy PE, Siriwardena G, Randle Z, Nowakowski M, Heard MS, Pywell RF (2022) The effects of a decade of agri-environment intervention in a lowland farm landscape on population trends of birds and butterflies. J Appl Ecol 2022. https://doi.org/10.1111/1365-2664.14246

Robinson DA, Hockley N, Cooper DM, Emmett BA, Keith AM, Lebron I, Reynolds B, Tipping E, Tye AM, Watts CW, Whalley WR, Black HIJ, Warren GP, Robinson JS (2013) Natural capital and ecosystem services, developing an appropriate soils framework as a basis for valuation. Soil Biol Biochem 57:1023–1033. https://doi.org/10.1016/j.soilbio.2012.09.008

Rodrigues FA, Bramley RGV, Gobbett DL (2015) Proximal soil sensing for precision agriculture: simultaneous use of electromagnetic induction and gamma radiometrics in contrasting soils. Geoderma 243:183–195

Sarkar R, Paul KB, Higgins TR (2019) Impacts of soil physicochemical properties and temporal-seasonal soil-environmental status on ground-penetrating radar response. Soil Sci Soc Am J 83(3):542–554

Schimmelpfennig D (2016) Farm profits and adoption of precision agriculture (No. 1477-2016-121190)

Schimmelpfennig D, Lowenberg-DeBoer J (2020) Farm types and precision agriculture adoption: crops, regions, soil variability, and farm size. Global Institute for Agri-Tech Economics Working Paper 01-20, https://doi.org/10.2139/ssrn.3689311

Sekulić A, Kilibarda M, Heuvelink GBM, Nikolic M, Bajat B (2020) Random forest spatial interpolation. Remote Sens 12(10):1687. https://doi.org/10.3390/rs12101687

Shafi U, Mumtaz R, García-Nieto J, Hassan SA, Zaidi SAR, Iqbal N (2019) Precision agriculture techniques and practices: from considerations to applications. Sensors 19(17):3796. https://doi.org/10.3390/s19173796

Sinfield JV, Fagerman D, Colic O (2010) Evaluation of sensing technologies for on-the-go detection of macro-nutrients in cultivated soils. Comput Electron Agric 70:1–18

Starks TH (1986) Determination of support in soil sampling. Math Geol 18:529–537

Stenberg B, Viscarra Rossel RA (2010) Diffuse reflectance spectroscopy for high-resolution soil sensing. In: Viscarra Rossel RA, McBratney AB, Minasny B (eds) Proximal soil sensing. Springer, New York, pp 29–47

Sudduth KA, Kitchen NR, Drummond ST (1998) Soil conductivity sensing on claypan soils: comparison of electromagnetic induction and direct methods. In: Robert PC (ed) Proceedings of the 4th international conference on precision agriculture, pp 979–990

Sudduth KA, Kitchen NR, Bollero GA, Bullock DG, Wiebold WJ (2003) Comparison of electromagnetic induction and direct sensing of soil electrical conductivity. Agron J 95:472–482

Sudduth KA, Kitchen NR, Wiebold WJ, Batchelor WD, Bollero WA, Bullock DG, Clay DE, Palm HL, Pierce FJ, Schuler RT, Thelen KD (2005) Relating apparent electrical conductivity to soil properties across the North-Central USA. Comput Electron Agric 46:263–283

Towett EK, Shepherd KD, Sila A, Aynekulu E, Cadisch G (2015) Mid-infrared and total x-ray fluorescence spectroscopy complementarity for assessment of soil properties. Soil Sc Soc of Am J 79(5):1375–1385

Tsan M, Totapally S, Hailu M, Addom BK (2019) The digitalisation of African agriculture report 2018–2019. CTA/Dalberg Advisers, Wageningen

Viscarra Rossel RA, Bouma J (2016) Soil sensing: a new paradigm for agriculture. Agric Syst 148: 71–74. https://doi.org/10.1016/j.agsy.2016.07.001

Viscarra Rossel RA, Walvoort DJJ, McBratney AB, Janik LJ, Skjemstad JO (2006) Visible, near infrared, mid infrared or combined diffuse reflectance spectroscopy for simultaneous assessment of various soil properties. Geoderma 131:59–75

Viscarra Rossel R, Adamchuk VI, Sudduth KA, McKenzie NJ, Lobsey C (2011) Proximal soil sensing: an effective approach for soil measurements in space and time. Adv in Agron 113:243–291

Wadoux AM, Marchant BP, Lark RM (2019) Efficient sampling for geostatistical surveys. Eur J Soil Sci 2019(70):975–989. https://doi.org/10.1111/ejss.12797

Webster R, Oliver MA (1992) Sample adequately to estimate variograms of soil properties. J of Soil Sc 43(1):177–192

Webster R, Oliver MA (2007) Geostatistics for environmental scientists, 2nd edn. Wiley, Chichester

Weindorf DC, Zhu Y, Chakraborty S, Bakr N, Huang B (2012) Use of portable x-ray fluorescence spectrometry for environmental quality assessment of peri-urban agriculture. Environ Monit Assess 184:217e227. https://doi.org/10.1007/s10661-011-1961-6

Wetzels H (2021) CAP beyond the EU – precision agriculture: for whom precisely? Article produced in cooperation with the Heinrich-Böll-Stiftung European Union: CAP beyond the EU | Precision agriculture: for whom precisely? | Heinrich Böll Stiftung | Brussels office – European Union (boell.org). Accessed 31 Aug 2022

Whelan B (2018) Site-specific crop management. In: Minasny B et al (eds) Pedometrics, pp 597–622

White K, Walden J, Drake N, Eckardt F, Settle J (1997) Mapping the iron oxide content of dune sands, Namib sand sea, Namibia, using Landsat thematic mapper data. Remote Sens Env 62:30–39

Yang X, Zhou X, Deng X (2022) Modelling farmers' adoption of low-carbon agricultural technology in Jianghan plain, China: an examination of the theory of planned behavior. Technol Forecast Soc Chang 180:121726. https://doi.org/10.1016/j.techfore.2022.121726

Zajícová K, Chuman T (2019) Application of ground penetrating radar methods in soil studies: a review. Geoderma 343:116–129. https://doi.org/10.1016/j.geoderma.2019.02.024

Zanella V, Ortiz BV, Thorp K, Morari F, Mosca G, Hoogenboom G (2015) Combining crop sensing and simulation modeling to assess within-field corn nitrogen stress. Precis Agric 15:391–398

Zhang Y, Hartemink AE (2020) Data fusion of vis–NIR and PXRF spectra to predict soil physical and chemical properties. Eur J Soil Sci 71(3):316–333

Žížala D, Minařík R, Skála J, Beitlerová H, Juřicová A, Reyes Rojas J, Penížek V, Zádorová T (2022) High-resolution agriculture soil property maps from digital soil mapping methods. Czech Republic Catena 212:106024

Chapter 3
Controlled Traffic Farming in Horticultural Systems

W. C. T. Chamen and J. E. McPhee

Abstract The aim of this chapter is to explore the concept of controlled traffic farming (CTF) as a means of confining soil compaction to the least possible area of horticultural fields, its positive outcomes and how it can be achieved. The chapter highlights the deleterious effects of soil compaction on the health of soils and productivity of crops. Excessive compaction has been caused by the increasing mass of farm machinery and its use in random patterns across the land. In addition to fully integrated CTF, a seasonal CTF system (sCFT) is also considered where not all machines can be integrated and robots can be used as an alternative to CTF for reducing soil compaction. Depending on the degree of confinement and machinery integration, CTF can improve soil porosity and permeability while reducing aggregate size and soil tare from harvested root crops. Access to the land can also be extended by virtue of the establishment of permanent traffic lanes.

Annual and perennial crop yields, both below and above ground, generally respond positively to the absence of traffic with responses thought to be associated with improved rooting and nutrient uptake. Remediation of compaction through subsoiling cannot be relied upon to sustain yield compared with CTF.

CTF can reduce energy inputs by up to 50% due to lower soil strength and easier seedbed generation, often reflected in fewer machinery passes. Consequently, with lower energy inputs and improved soil health, greenhouse gas and particularly nitrous oxide emissions can be substantially reduced (up to 44%). With lower energy inputs and greater yields, farm gross margins are generally improved with both CTF and sCTF, but their extent is governed by other costs, such as automatic steering systems.

Transitioning to CTF needs a whole systems approach that recognises the challenges of lower inputs and the preclusion of ploughing (unless a seasonal system is adopted). More diverse cropping systems increase the timescale for change and the

W. C. T. Chamen (✉)
CTF Europe, Shrewsbury, UK
e-mail: tim@ctfeurope.eu

J. E. McPhee
Tasmanian Institute of Agriculture, University of Tasmania, Burnie, Tasmania, Australia

P. Newell Price et al. (eds.), *Precision Agriculture in Field Horticulture*, Progress in Precision Agriculture, https://doi.org/10.1007/978-3-031-96534-0_3

likely costs of conversion, but both can be reduced by careful planning and lead to lower production costs. Short-term renting of land is an additional challenge but could be diminished by wider adoption of CTF.

The future use of gantries, which are initially a more costly transition, are also considered and the potential for such systems to overcome many of the shortcomings of existing machinery while adding many other advantages.

Overall, CTF integrated with very small autonomous vehicles, is considered the most practical means of addressing the present conflict between using large and heavy machines and improving soil health. The wider adoption of CTF will play a key role in supporting sustainable horticultural practice and particularly its future manifestation in the form of wide span or gantry tractors. The chapter includes two farm case studies that highlight aspects of the adoption process.

Keywords Controlled traffic farming · Farm machinery · Soil compaction · Crop yields · Soil health · Adoption · Sustainable practice

3.1 Introduction and Background to Controlled Traffic Farming (CTF)

It is generally accepted that humans started out as "hunter gatherers", gaining sufficient food to survive by learning through experience what plants were edible, where they grew and what time of year their leaves, fruits or roots were at their best for consumption. At some stage, humankind started to farm these crops, such that people were less at risk of shortage and could settle in one spot rather than being reliant on a nomadic existence. As farming developed, people moved from a situation where they had little or no impact on the way plants grew to controlling many aspects of a crop's growth. However, humanity has now reached a point in mechanisation where it's negative impacts are starting to limit the efficiency with which crops can be grown (Håkansson and Reeder 1994; Keller et al. 2019; Keller and Or 2022) and has led to a plateauing in crop yields (Keller et al. 2019) that are far from reaching their genetic potential. Even light machines used repeatedly can create significant compaction of the topsoil (Calleja-Huerta et al. 2023; Nawaz et al. 2023).

This chapter concentrates on the underlying causes of these problems and describes a means of addressing them together with suggestions for implementation as well as examples of on-farm adoption. Primarily, the chapter focusses on annual, biennial or multi-annual crops that are traditionally associated with extensive tillage. Although perennial crops, such as vines, coffee beans, kiwi and many other fruits produced in orchards, face similar issues (Lanyon et al. 2004), they are outside the remit of this chapter.

The Problem with Traffic

One of the factors limiting crop yields is uncontrolled soil compaction arising from the wheels or tracks supporting agricultural machinery. This is primarily associated with seed- or root-bed preparation and harvesting operations. Growers cultivate to bury crop residues and weeds, while also remediating soil compaction (caused by harvesting operations) and creating an appropriate tilth into which new crops can be sown or planted. Although 'soil compaction' simply describes an increase in the dry bulk density of a soil, it also leads to low porosity, reduced aeration and water infiltration, increased soil loss through erosion and elevated soil strength that impacts negatively on seedbed preparations (Chamen et al. 1990, 1992; Esteban et al. 2019; Shaheb et al. 2021).

Most efforts to create an "ideal" seed and root bed involve alleviating the compaction caused by the equipment used for the previous crop and to a lesser extent, manipulating conditions so that subsequent machinery operations will not be compromised. Whatever is done in this respect does not lead to a consistent outcome because the initial soil conditions, in any given field and any given year, are wildly variable; some soil having been trafficked once, other areas twice or more and some areas not at all. The interest in and expansion of the area under no-till (Derpsch et al. 2010) rather reflects an admission that doing nothing to the soil can be just as productive as deep and rigorous tillage. The fact is that it is not really known exactly what soil condition is ideal for different crops and it is unlikely to be found out or optimally achieved if soils continue to be trafficked at random. This creates additional variability, which is a further complication to precision farming (Vermeulen et al. 2010; Antille et al. 2019a, b; Chamen et al. 2025). Until a production system that avoids the inconsistency caused by random trafficking is embraced, it will be difficult to understand and improve soils for the objective of optimum crop production. Fortunately, CTF is a solution to this problem that is conceptually simple, although its ease of application varies widely depending on industry and location (Tamirat et al. 2022).

Controlled Traffic Farming

Controlled traffic farming, or CTF as it has become widely known (Misiewicz and Galambošová, 2023), addresses inconsistencies in the soil condition caused by random trafficking by dividing the land into tracked and non-tracked zones. The basic philosophy of CTF is to achieve as much non-trafficked soil as possible within the constraints of cost and machinery capabilities. CTF provides the ability to loosen or firm, cultivate or not and as a result, forms the basis of a more efficient means of farming, and particularly precision farming. If non-trafficked soils cannot be made to yield significantly more, it is just that the way to do it has yet to be learnt!

An example of the disconnection between precision on the one hand and ineffectual soil management on the other was provided during the author's observation of a 12 m wide inter-row hoe, accurately guided by a vision guidance system in a cereal crop. The condition of the Evesham series clay soil following traditional management was such that the hoe would either not penetrate, or when it did, resulted in the break-out of troublesome clods. Invariably, a non-trafficked soil will present conditions that are far more amenable and consistent, allowing a hoe, for example, to penetrate a more friable medium rather than an intractable mass, a fact observed by Halkett in the 1850s when traffic was primarily of an animal nature (Halkett 1858).

This chapter introduces the principles of CTF within a field-based horticultural context. It describes the advantages that it can bring, and how CTF can be achieved within the constraints of existing machinery. It then considers how widely CTF has been adopted and finally, provides examples of systems that are working in practice.

Most importantly, mention is made of gantry tractors that remove most of the constraints that existing machinery places on CTF efficiency and effectiveness.

Principles and Practical Implications of CTF

The underlying principle of CTF is to maximise the area of non-trafficked soil by confining all vehicle traffic to the least possible area of permanent traffic lanes. Accepting that it is not presently possible to avoid vehicles running on the soil altogether, then minimising the areal extent of their impact is the next best thing. The areal extent relies heavily on what crops are being grown and the dimensional characteristics of the machinery used to manage them. Minimising the tracked area involves matching up as closely as possible the track gauges (centre distance between wheels on an axle) on all field vehicles and exactly matching their implement operating widths (Isbister et al. 2013). Direct multiples of the operating width can also be used, such as with sprayers and fertiliser spreaders.

Minimising tracked area also requires that machines are steered very precisely on permanent traffic lanes, which is normally facilitated by global navigation satellite systems (GNSS). These, with the help of a ground- or satellite-based correction signal (real-time kinematic (RTK)) and auto-steer can keep vehicles to within ±20 to ±30 mm of their intended path, giving greater accuracy than can be achieved manually (Carballido et al. 2014; Yigit et al. 2022). These automated systems (which are widely used without adoption of CTF) perform steering corrections and allow the driver to concentrate on machine operation. These systems can be delivered onto farms in many ways, but their detail (other than as a cost) is not considered in this chapter.

CTF should lead to a significant reduction in tracked area compared with current practice, where the land can be tracked multiple times during the growing season of any horticultural crop (Domzal et al. 1991; Kuipers et al. 1994). Although increased machinery capacity in recent years has reduced the cumulative area of wheel tracking, many horticultural crops still experience multiple passes over the same area during a growing season, with a large proportion of the coverage occurring at harvest (McPhee, unpublished data). In arable farming, research by Kroulik et al. (2009) showed that even with no-till (sowing or planting of crops with no prior cultivation), around 45% of the soil is run on by wheels during crop establishment, rising to around 85% for plough-based systems. In addition, the location of these areas will vary from season to season, meaning that all the soil will have been trafficked at least once within three seasons. In terms of the vegetable industry, McPhee and Aird (2013) point out that the working width and track gauge of most horticultural machines are incompatible, making it far more difficult to achieve low-tracked areas. Recognising that the standardisation of track gauges and working widths is still some way off, they encouraged the adoption of seasonal CTF (sCTF) or a hybrid system, with a narrower track gauge of perhaps 2–2.5 m for horticultural crops and a wider gauge of 3 m where grain crops are part of the rotational mix. Track gauges as wide as 3.2 m are widely used for horticultural crops in the Netherlands with farmers accepting that these machines are generally too wide to be used on public roads. Accepting this constraint, a farmer in the UK has extended his combine harvester track gauge to 3.86 m so that his narrower gauge machines can run at 1.93 m, exactly half the wider gauge. This allows the grower to plant potatoes on soil that has not been trafficked for some years in a cereals rotation before being compromised by incompatible machinery during potato harvest. Other growers (see Case Study 3.2) however, have modified their potato harvester's track gauge to match that of their CTF system, so that they are able to maintain non-trafficked beds through their potato harvest.

In horticultural systems the row spacing of a particular crop can have a major impact on crop density with different track gauges and tyre widths. In some cases, negative influences of a CTF system can be counteracted by sowing or planting more densely alongside the traffic lanes where more light is available.

Thus, although the principle of CTF is simple, achieving it in practice with a diverse and non-standardised range of machinery can be challenging. The advantages of CTF need to be substantial to warrant change when the barriers to adoption are significant.

The Australian Controlled Traffic Farming Association (ACTFA) state that three principal requirements need to be met for a successful CTF operation (https://www.actfa.net), namely:

- common track gauges and implement widths;
- satellite guidance and auto-steer using an RTK correction;
- farm planning for erosion control, drainage, irrigation, crop husbandry and field logistics.

Meeting these requirements will deliver three key benefits:

- elimination of most, if not all, tillage operations (Chamen et al. 1990)
- reduced fuel use and tractor time (Vermeulen et al. 2010; Chamen and Longstaff 1995)
- reduced investment in tractor power and tillage equipment (Chamen and Audsley 1993)

3.2 Benefits of CTF

Fully Integrated CTF Systems

The benefits of controlled traffic extend across an entire cropping system and can include improved soil structure, biology, infiltration, soil aeration, soil water storage, internal drainage, water and fertiliser use efficiency, crop growth and yield, timeliness and economics as well as reduced soil tare with root crop harvests (Antille et al. 2019a, b; Dickson et al. 1992; Kingwell and Fuchsbichler 2011; McGarry 2003; McHugh et al. 2009; McPhee et al. 1995b, 2015; Neale and Tullberg 1996; Powrie and Bloomer 2010; Rodgers et al. 2018; Vermeulen et al. 2010; Kaczorowska-Dolowy et al. 2019; McPhee 2020; Hussein et al. 2021; Godwin et al. 2022; da Luz et al. 2023). The benefits of CTF are also measured in reductions of negative aspects and costs, such as reduced runoff, erosion, soil borne diseases, energy use for tillage, greenhouse gas (GHG) emissions, number of tillage operations, tillage equipment inventory, tractor size, capital and operating cost and operating hours (McPhee et al. 1995a, 2015; Neilsen 2008; Stirling 2008; Tullberg et al. 2018; Tullberg 2000). Most of the research that has quantified these benefits has occurred in the arable cropping sector, but it is likely that many of these benefits will be magnified in horticultural cropping systems where soil working is normally more intensive. As mentioned in the previous paragraph, research on controlled traffic in vegetable and root crops has been limited, but there have been a few key studies. These studies are listed and described below before quoting their results (with authors in italics) in terms of crop performance, soil conditions, greenhouse gas emissions and energy and costs.

Key Studies

Lamers et al. (1986). In this 8-year study at two centres in the Netherlands, sugar beet, potatoes and winter wheat were grown on a loam soil under a full CTF system using machines with a common 3.3 m track gauge. A rotary digger was used at 0.2 m depth on the CTF plots while mouldboard ploughing and fixed tine cultivation to 0.25 m and 0.28 m depth respectively were used on the conventional traffic plots. All field operations were conducted on the same date.

Harrison et al. (1985). This study was more aligned to vegetable cropping and was conducted at Luddington in the UK. Their work involved ploughing a sandy clay loam (with and without pebbles) and then imposing wheeling on the following percentage areas with a tractor weighing 2080 kg on a 1.83 m track gauge:

zero,
50% once,
50% twice,
90% once at right angles to the direction of ploughing.

The beds formed were then power-harrowed to around 110 mm depth using the fixed traffic lanes, which were also used for all other operations. Crops grown were lettuce (*Lactuca sativa*), bulb onions (*Allium cepa L.*), parsnips (*Pastinaca sativa*), carrots (*Daucus carota*) and cauliflower (*Brassica oleracea*).

Vermeulen and Mosquera (2009) grew green peas (*Lathyrus oleraceus*), spinach (*Spinacia oleracea*), onions (*Allium cepa* L) and carrots (*Daucus carota*) on a loam soil on an organic farm in the Netherlands. However, unlike Lamers et al. (1986) they were unable to sustain full controlled traffic for harvesting and post-harvest primary tillage (ploughing) and thus named it "seasonal" controlled traffic (sCTF), which they compared with conventional random traffic farming.

Hefner et al. (2019) conducted their CTF research on a fine sandy loam and a coarse sand on two organic vegetable farms in Denmark. The CTF system tracked 30–40% of the area (primarily associated with the harvesting of beetroot (*Beta vulgaris*) and potatoes (*Solanum tuberosum*) while tillage was confined to 250 mm depth with a stubble cultivator. At the first site, each of the fields was divided into two plots (one CTF, the other random traffic (RTF)), with a further subdivision for five different crops within each year, while at the second site only one crop was grown each year.

Pedersen (2015) carried out compaction trials to determine the effectiveness of subsoiling as an amelioration measure on a commercial farm growing onions that had employed sCTF for the previous 5 years. Treatments were compaction or no compaction, each with and without subsoiling. The soil was a loamy coarse sand and a crossed split plot experimental design with four replicates was used. Compaction was applied on first November using a fully laden potato harvester mimicking typical harvest traffic intensity so that the whole area was covered at least once by a tractor or harvester wheel. This was immediately followed on the subsoiled plots with deep loosening to 400 mm depth, but without further tracking in the growing bed for this and for subsequent tillage and planting operations on these and any other plots. Prior to these treatments in November, the site had been cropped with potatoes in May, subsoiled to a depth of 500 mm in June and planted with a cover crop using controlled traffic.

Pedersen (2014) took a model based on five Tasmanian vegetable farms (growing, for example, broccoli (*Brassica oleracea var. italica*), carrots, onions, peas, poppies (*Papaver somniferum*) and potatoes), and revised it for use with a wide span vehicle to simulate greenhouse gas emissions in relation to gross margins of conventional and different CTF systems. The conventional system assumed nitrous oxide emissions from soil of 51%, 18% from irrigation, 12% from fuel and 8% from pesticide applications.

Effects of CTF on Soil-Related Conditions (Based on the 'Key Studies')

Results from *Lamers* et al. revealed that tillage draught was reduced by 25% in the absence of traffic while the traffic lanes provided a 13% improvement in tractive efficiency and the system overall resulted in an energy saving of 50%. Over the

8-year period of the experiment, soil porosity improved from 51% to 55% (vv^{-1}) while saturated water permeability in the 0.12 to 0.17 m layer increased to 3 m day^{-1}, compared with 1.5 m day^{-1} on the conventional plots. In the same layer, oxygen diffusion doubled in the absence of traffic, ensuring that oxygen supply to developing roots was always sufficient. Measures of aggregate size distribution in the potato seedbed showed smaller aggregates on the controlled traffic plots, resulting in a 27% (ww^{-1}) reduction in soil tare. They also made the point that for shallow sown crops the permanent traffic lanes enable sowing to proceed earlier because the soil in the crop bed only had to be workable rather than trafficable. Few problems were reported other than slippery traffic lanes in very wet conditions and the inability of fixed tine tillage to counteract the effects of soil slaking after potatoes. The latter was overcome by using a rotary digger at the lowest possible rotor speed.

Vermeulen and Mosquera found that despite the shortcomings of a seasonal system they were able to demonstrate improvements in topsoil structure (pore space, water permeability, aggregate size distribution) with sCTF for the crops grown on the flat, but not for carrots that were grown on ridges. Soil structure for all crops, including carrots, seemed to have improved on this farm together with yields, which were well above average (Bernaerts, personal communication, 2022).

Pedersen (2015) found that the compaction treatment increased the draught force of the subsoiler by 28%, (from 34.6 to 44.3 kN) while bulk density was reduced by the loosening operation but only significantly compared with the plots that had not been loosened following the compaction treatment. As might be expected, macropores in the loosened but not compacted plots were significantly greater than those in the compacted but not loosened soil to 250 mm depth, as were those in the compacted and loosened soil at 250 mm depth. Other differences were apparent, but not significant. Penetration resistance followed a similar trend to the macropore data.

Effects of CTF on Crop Performances (Based on the 'Key Studies')

Lamers et al. reported modest yield increases compared with conventional traffic ranging from 3% to 7% in potatoes, up to 6% in sugar beet (*Beta vulgaris*) (particularly following a wet June or July) and 7% in wheat.

Harrison et al. found that although effects varied with soil type and soil moisture content, compaction on the pebbly soil reduced the total yield of lettuce in two out of three years, the yield of parsnips in one year out of two (due to fanging) and the marketable yield of bulb onions in one year. Curd size in autumn cauliflower was also reduced on the pebbly soil but only significantly where 90% of the area was compacted. There were no compaction effects on carrots, possibly due to drier conditions when the wheelings were imposed.

Vermeulen & Mosquera discovered that crop yields increased significantly under CTF for green pea, spinach and planted onion sets but not for carrots and direct sown onions.

Hefner et al. found that in the third year after CTF introduction, the marketable yield of white cabbage (*Brassica oleracea var. capitata f. alba*), beetroot and potatoes increased by 25%, 40% and 68% respectively on the fine sandy loam under CTF compared with conventional practice. However, none of the differences in the yield on either soil were significant in the following year (2016), although root growth was greater under CTF. Furthermore, spring soil mineral N was similar or higher under CTF compared with RTF on the fine sandy loam and was thought to be associated with higher crop yields and residues under CTF. Overall, responses to CTF were positive and the authors concluded that CTF had the potential to increase vegetable yields and root growth while increasing N supply compared with RTF.

Pedersen found that CTF improved onion yields compared with RTF, despite compaction remediation on the trafficked plots appearing to be successful as indicated by physical measurements. The marketable yield was greatest where there was neither compaction nor subsoiling, suggesting that a reduction in tillage intensity was perfectly feasible for maintaining or improving onion production. Root proliferation was always greater following subsoiling but where there was no loosening, the non-trafficked plots had greater root growth at 0.2 and 0.3 m depth, but no difference at 0.1 m compared with RTF. There was a negative correlation between penetration resistance and root growth ($P < 0.05$) but no correlation with yield. Although the effect was not significant, *Pedersen* noted that the treatments resulting in the lowest penetration resistance and greatest root intensity (i.e., non-compacted and loosened) did not necessarily result in the greatest marketable yield.

More recent research (van Balen et al. 2023, but not one of the key studies) investigated the effects of reduced tillage (with and without subsoiling) in a CTF scenario compared with conventional and organic farming plough-based systems. They grew 13 different crops over a 10-year period, including carrots, potatoes, sugar beet, cabbage (*Brassica oleracea*), onions and cereals, most of which sustained their marketable yield compared with the conventional and organic systems. The exceptions were carrot and cabbage, the former possibly being negatively affected by a preceding cover crop and larger aggregates compared with the other systems that were ploughed in the autumn. Reduction in cabbage yields (organic rotation without subsoiling) were primarily associated with one year when the preceding grass-clover cover crop was poorly terminated.

In terms of perennial production systems, such as those used for vines, coffee beans (*Coffea*), kiwi (*Actinidia deliciosa*) and many other fruits produced in orchards, soil compaction can be equally negative as far as crop yields are concerned. Vines in particular, where inter-row above-ground operations place vehicle wheels over the rooting zone, can experience negative effects on crop yields but also on water run-off and soil erosion (Lanyon et al. 2004).

Effects of CTF on Nutrient Use efficiency and Greenhouse Gas emissions (Based on the 'Key Studies')

Vermeulen and Mosquera reported that although available N-min at the end of the cropping period did not differ between treatments, the combination of 20–50% lower nitrous oxide emissions and improved yields suggested that N-use may have been more efficient under the sCTF system.

Pedersen's research (Pedersen 2014) predicted an average 31% reduction in greenhouse gas emissions for a range of CTF systems compared with conventional practice with a 16% reduction for sCTF, but as much as 44% for a dedicated wide span system. Most of the reductions were through lowered nitrous oxide emissions from the soil.

Effects of CTF on Energy and Costs

Vegetable production tends to rely heavily on tillage for the amelioration of harvest traffic compaction and the establishment of seedbeds. The isolation of soil compaction to permanent traffic lanes can make a significant difference to the tillage required for seedbed preparation, and consequently to the tractor power needed on a farm. As noted previously, Lamers et al. reported a 50% reduction in energy use under controlled traffic conditions. Dickson and Ritchie (1996) reported 40–50% increases in tillage power requirements for random, compared to non-trafficked soil in an arable rotation that included potatoes in Scotland. Data from three research and commercial sites in Tasmania, Australia, showed reductions in the number of tillage operations of 20–60% compared to systems operating under conventional traffic (McPhee et al. 2015). Although no draught or power consumption measurements were taken, this change in the number of passes represents a major reduction in energy investment in cultivation. This saving in energy and improved seedbed conditions under CTF was also evidenced by Chamen et al. (1992), who revealed

Fig. 3.1 Contrast in tilth following ploughing on a clay soil that had been randomly trafficked (left) compared with an area which had been maintained traffic-free (right)

a 60% reduction in specific plough draught alongside a much finer tilth on non-trafficked compared with conventionally trafficked clay soil (Fig. 3.1). These contrasts were reflected in the total energy required to create a spring barley seedbed following autumn ploughing, which was just 79 MJ ha^{-1} with CTF compared with 255 MJ ha^{-1} for traditional practice.

Vermeulen & Mosquera concluded that sCTF was more profitable than random traffic farming (RTF) if the average crop yield increased by more than 1.6% on a 50-ha farm and by 2.2% on a 200-ha farm, despite machinery costs for sCTF being higher. As the average yield increase in their study was 6–10%, they considered that sCTF was an economically viable option in organic vegetable farming, even without accounting for any timeliness benefits, which were anticipated by the authors.

The model developed by *Pedersen* predicted a 59% increase in average farm returns for CTF systems compared with conventional practice (McPhee and Pedersen 2017). This was reflected in gross margins that were 7% higher for sCTF and 44% higher for a wide-span system, but benefits were not universal to all farms and crops. In particular, the economic gains associated with peas and autumn-planted broccoli were outweighed by the cost of automatic steering systems.

Seasonal Controlled Traffic Farming

As CTF has only been practised widely for a few decades and has still to achieve mainstream use, development of customised machinery for its implementation has been slow, particularly in the horticulture sector. As a result, farmers, rather than having a fully integrated CTF system with all machinery matching, have often had to accept a "halfway house" or what has become known (and mentioned earlier) as a seasonal CTF (sCTF) system. A seasonal CTF system is one where for one or more operations, the farmer has to abandon the permanent traffic lanes because the machinery is incompatible. In practice, the latter generally relates to harvest machinery meaning that there can be extensive "out of track" compaction, which must be ameliorated by deep loosening or ploughing. This loosening is also generally undertaken on a conventional random traffic basis, but more benefits can be retained if it is conducted by driving on the permanent traffic lane locations or, if ploughing, that this is undertaken with the tractor "on the land" rather than in the furrow. The latter will avoid compaction deeper in the soil profile, which is persistent and costly to remove. This is particularly important on lighter soils (sands and silts) that can develop an impervious pan at plough depth but equally on clays that often remain moist at plough depth.

Chamen and Longstaff (1995) researched the effect of a sCTF system within an arable rotation where conventional equipment was used for harvesting and ploughing but all other operations were confined to permanent traffic lanes, i.e., there was no further traffic on the cropped soil. This treatment was compared with conventional practice and with a full CTF system, all systems having been in place for around six years and using a plough for primary tillage. Measurements were

Table 3.1 Tillage (ploughing) energy, soil tilth and crop yield on an Evesham series 60% clay soil managed with different traffic treatments

Traffic treatment	Forward speed, ms^{-1}	Depth of operation, mm	Draught per unit width, kN	Specific energy, kJ m^{-1} width	Tilth, mean weight diameters*	Yield of wheat, t ha^{-1}
Conventional	1.05	191	7.8	13.55^a	18.1	8.55^c
CTF	0.96	207	5.4	8.55^b	16.7	10.01^d
sCTF	1.03	173	7.6	13.96^a	24.4	9.49^c

*Measure of the "coarseness" of a tilth. The higher the number, the coarser or more cloddy the tilth
Values with different letters significantly different at: ^{a,b} $P < 0.001$; ^{c,d} $P < 0.05$

Fig. 3.2 Example of a seasonal CTF system on an organic farm in the Netherlands. Carrot harvest, followed by ploughing, used conventional methods without traffic control. This shows the permanent traffic lanes being re-established while cultivating for the next crop and evidence of the compaction caused by the harvesting operations

made of tillage energy, soil tilth and crop yield. Table 3.1 provides an overview of the results.

From these data it is clear that operating a sCTF system may only be marginally better than conventional practice, although there was a non-significant but 11% increase in yield. This is borne out by visual evidence from an organic farm in the Netherlands that was unable to maintain their controlled traffic operations during the harvest of carrots in very wet conditions. Figure 3.2 shows the soil post-ploughing on this silty soil, some of which had become anaerobic. This farm now uses a shallow plough operated from the traffic lanes, meaning that the compaction created during the ploughing operation can be avoided (Bernaerts, personal communication, 2022).

3.3 Understanding the Implications of and Adapting to the Benefits of CTF

CTF provides a blueprint for permanently limiting the area of compaction on field soils. But CTF is about far more than just confining traffic—it is about changing farming systems to respond to the more amenable and improved soil conditions and to greater precision in operations. Under CTF, farmers no longer have to manage a cycle of soil damage and repair; most soils maintain a good healthy structure once the damaging effects of wheels have been removed and they have been given time to ameliorate (Chamen 2011). If subsequent amelioration is required, this will need less energy and result in a more favourable condition than was formerly the case, as illustrated in Fig. 3.3.

Depth of working can also be lessened as soil structure improves (McHugh et al. 2009) and there is no further damage at depth due to heavy loads. However, there are challenges. Traditional systems based on ploughing and power-intensive tillage operations, such as rotary hoeing and power harrowing, leave the soil surface clear of weeds and residues, a condition that is often critical to the reliable and subsequent operation of precision horticultural transplanters, planters and drills. With current technology it is very difficult to plough and maintain a CTF regime (but see section on gantry tractors), so new skills, procedures and precision are required, such as residue management to enable minimum or strip-till or direct planting to replace the residue-free surface left within a ploughing system.

Fig. 3.3 Soil conditions in the same field and soil type following deep loosening on conventionally managed soil (left) and an area managed with CTF (right) showing the overall finer tilth that has been maintained

As noted, soil conditions under a non-trafficked regime are very different from those under conventional practice. Under CTF, growers are more likely to ask: "why do I need to do any "re-structuring" tillage before planting the next crop?" However, some soils will naturally settle or slump even without imposed compaction, particularly if there has been a history of rigorous tillage and organic matter levels are low. CTF can't wave a magic wand to cure this problem, but it should set growers on a path of reduced or zero tillage with greater potential to maintain ground cover throughout the year. This will gradually elevate organic matter levels, which in turn will help stabilise the soil. As soils recover, growers should be able to ask: "has the soil condition improved sufficiently for it to no longer impair soil function (aeration and drainage) without tillage or the yield and quality of the crop?" For example, Pedersen (2015) found that despite a higher penetration resistance on non-trafficked compared with trafficked and loosened soil, yields were still greater under the non-trafficked regime. Using dry bulk density and arthropod abundance as measures of soil physical condition, McPhee et al. (2015) and Rodgers et al. (2018) found that non-trafficked soil exhibited more desirable characteristics than soil exposed to random traffic, even though both traffic conditions were subjected to conventional, disruptive tillage operations. Similar observations were made by Lamers et al. (1986).

In the early years, following a transition to CTF, this question of soil "quality" is likely to be asked more often and will continue until more carbon is sequestered and the soil has stabilised. Additional carbon in the form of organic matter helps give any soil more resilience, in essence a more 'elastic' rather than 'plastic' behaviour where changes due to imposed loads are reversed over time. An example of increased resilience on a swelling clay soil was demonstrated by Chamen (2011) when a range of heavy machines were driven over land that had been in controlled traffic for five years. Within approximately nine months the significant increase in soil strength caused by these single passes of machinery had largely been dissipated, as indicated by a reduction in penetration resistance (Fig. 3.4). This contrasted with the data reported by McHugh et al. (2009), whose research considered recovery of a randomly trafficked soil when traffic was removed. In this situation, recovery, particularly deeper in the profile, was measured in years rather than months.

Controlled traffic farming has two elements to it, firstly a purely mechanical element that involves matching up all the machinery to minimise the tracked area, and a second more challenging element that involves changing the farming system to maximise the benefits that traffic segregation can bring. As has been pointed out above, without changes in tillage for example, one would not be able to secure all the benefits that controlled traffic farming can bring, but changing tillage also changes the agronomy. The recent research by van Balen et al. (2023) demonstrates that minimum tillage or conservation agriculture for root, planted and small seeded crops under a CTF regime can be just as successful as conventional practice provided crop residues are dealt with effectively. Some farmers argue that introducing no-till or conservation agriculture without also switching to CTF is not viable and *vice versa*.

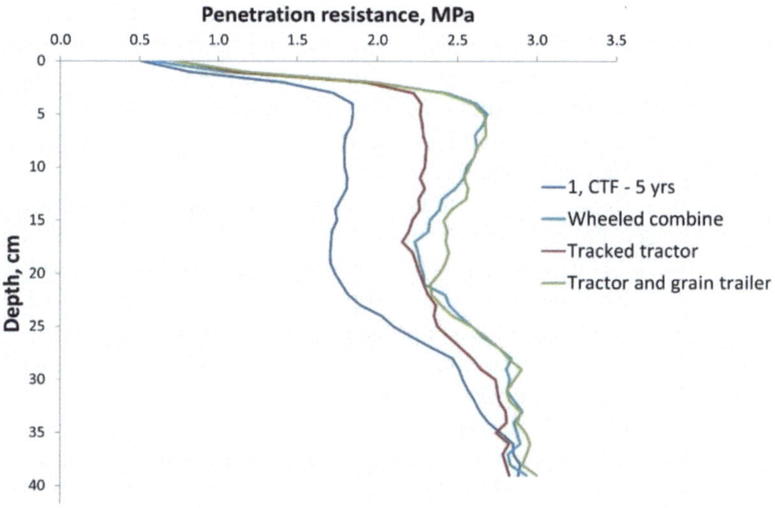

Fig. 3.4 Cone penetration resistance on a clay soil which had been in controlled traffic for five years showing the effect of a single pass of various machines on post-harvest stubble

Reducing Tillage Inputs

Adopting CTF within high-value cropping systems without significantly reducing tillage inputs is missing one of the key benefits of traffic control. Only by reducing tillage can one start to realise a reduction in machinery investment and production costs as well as an improvement in soil health. However, most agricultural production systems are fragile and only work reliably in a limited range of field conditions. Thus, soil moisture content can be too high or too low, while some cultivators or drills will only work satisfactorily with limited surface residues, and often only when these too are in a certain state. The key challenge with CTF is to be able to reduce the depth and intensity of tillage in such a way that the reliability of any operation is not diminished, and preferably is improved.

An example of adopting CTF into a high-value crop is a traditionally managed farm growing onions that invariably autumn ploughs following a cereal crop in preparation for spring planting or sowing. In the spring, the ploughed land is often rigorously cultivated to break up and level the tilth before the drill or planter will work satisfactorily. Although some of these operations will have compromised the soil deeper in the profile, drills or planters are likely to operate reliably in what is a relatively level surface free of weeds and residues. If the land is being rented from a non-CTF grower, there will be little alternative to maintaining this system in the first year. However, where the land (owned or otherwise) has been in CTF for a preceding cereal crop, there should be potential for reducing the intensity of tillage before an onion crop for example. Autumn ploughing could be replaced by shallow tillage, but only if chemicals applied to the previous crop aren't an issue and don't need to be buried. This alternative cultivation will almost certainly need to be repeated in the spring to further incorporate residues and to counteract any slumping and to aid drying of the soil. If surface conditions have been rendered favourable, the onion crop could then be sown or planted knowing that the surface and sub-surface profile have remained undamaged by traffic.

Reducing tillage inputs is likely to be an iterative or 'trial and error' process but until the grower gets the process right, ploughing may still be needed within a CTF regime. This should preferably be practised by running the tractor on the soil surface rather than in the furrow and with the lowest possible tyre pressures. It could also be undertaken at a shallower depth. As indicated earlier, the soil will turn over in a more amenable state following a period in CTF, allowing a seedbed to be created with lower inputs (Chamen et al. 1992).

Adopting strip tillage is another means of reducing inputs and will be less challenging within a CTF regime compared with conventional practice, where contrasts in soil compaction often lead to varying quality of soil tilth. As with shallow tillage, an autumn pass for spring crops should be "freshened up" a short while before planting.

Introducing No-Till or Direct Planting

The ultimate in reducing tillage is to direct plant and this is becoming increasingly feasible with new designs of drill, but is still challenging for closer-spaced row crops such as onions and carrots. However, as with conventional planting, opposite traverses of each bed using a planter with wider rows and an appropriate offset can provide the narrower row spacing required. An alternative is to have a planter with staggered rows (two sets of units on separate toolbars one behind the other) but this results in heavier machines and a centre of gravity further behind the tractor, which may need to be counteracted by additional weight on the front of the tractor. Alternatively, a trailed unit could be used, but this would probably need another guidance and steering system to keep the drill on the correct drilling line. Key to the success of no-till is a healthy soil and it may be some years into a full CTF system before suitable soil conditions eventuate, so practitioners should always be inspecting their soils to determine the feasibility of a no-till approach.

3.4 Putting CTF into Practice

Full CTF

Horticultural systems are far more complex than those associated with arable production, where most crops are annual. In field horticulture, farms often undertake short-term rentals of land to extend the period between growing a particular crop on the same land. The implications of the latter in terms of CTF are profound. Does the farmer invest in setting up a complete CTF system on land that might only be under his or her control for one or two years? In this situation, the answer will revolve around the level of investment needed and the likely return over what may only be a short period of time (see Case Study 3.1 for information about transition costs).

From a CTF perspective, fully mechanised production systems tend to be more complex than those where hand-harvesting or simple mechanisation can be employed. The latter generally use tractor-based systems on single or multiple beds and are often associated with crops such as lettuce, brassicas, baby-leaf lettuce or sweetcorn (*Zea mays convar. Saccharata var. rugosa*). The track gauges of many machines are still based on the imperial size of 6 feet (1.83 m), and with many machines having tyres just 320 mm wide, trafficked areas can be as low as 18%. Multiples of this gauge for tillage machines lead to a wide variety of widths but farmers are often happy to modify machines to fit, or to order customised versions. Equally, some systems are based on a track gauge of 1.5 m, which if used with machines of 1.5, 3, 4.5 or 6 m width can provide trafficked areas of anywhere between 11% and 21% depending on the tyres used. At the other extreme is a track gauge of around 3.2 m, resulting in a trafficked area of around 12%, which is compatible with machines of 3 m working width if the wheel tracks are not being worked. In between these two extremes is a gauge of 2 m, which is becoming more common on present-day machines (particularly on arable farms), but plant row spacings would need to change to fit into a new bed width.

There is also the issue of machine choice, which is limited for harvesting beans and peas for example, and very few are presently compatible with CTF systems. Additionally, there are the loads involved, which are far greater than with arable systems (potatoes for example) and emphasise the need to minimise pressures by not constraining the width of tyres beyond practical bounds. Running a CTF system also means that drivers cannot take a "short cut" to the field exit or vice-versa when entering the field to go to the harvester. Bochtis et al. (2010) researched the effect of this constraint by comparing a conventional system with a CTF system when applying slurry using an application vehicle supported by a refilling unit, i.e. similar to a harvesting operation but in reverse. First, they measured the actual performance of the two systems in two different fields and then simulated each in the opposite field. On average they determined that the CTF system was about 5% less efficient than the conventional system and resulted in about 36% extra travel distance, albeit on firm traffic lanes.

Deciding on a system is therefore dependent upon many factors and all aspects of cropping must be considered well into the future to ensure compatibility.

CTF will invariably make organic production easier as the precision associated with mechanical operations is of critical importance. Hoeing, for example, as mentioned earlier, will not only be less disrupted by large contrasts in soil strength, but also more effective because weeds are more likely to be pulled out rather than cut off at the soil surface. Timeliness of operations should also be improved through better access and free-draining soil conditions. Reduced fuel and power demands are highlighted in Case Study 3.1.

There is presently no single model for achieving full CTF within a horticultural cropping system. Every farm must scrutinise its existing machinery to determine the least cost approach to conversion. The two key decisions are; (a) what common implement width and (b) what vehicle track gauge to adopt. Once these elements have been determined, any given farm can plot its way to achieving a complete CTF system providing it doesn't have a plethora of incompatible harvesters or rely heavily on contractors. In these latter cases, only greater demand for CTF-compatible machines will bring about change, and this is likely to be very gradual. Economics will dictate the time scale for adoption because most farms can only afford to make changes within the normal machinery replacement cycle. Case Study 3.1 illustrates a typical process of change when transitioning to a CTF system.

Seasonal CTF

Electing to use sCTF, as mentioned earlier, usually comes about because the working widths and/or track gauges of harvest machinery are incompatible and not easily changed to match up with other machines in the CTF system. The CTF element therefore usually consists of matching up the secondary tillage and drilling/planting machinery as well as any implements used for in-crop operations, such as inter-row hoeing and chemical applications. Primary tillage (ploughing or non-inversion deeper loosening) may or may not be conducted from permanent tracks depending on the width of operation.

Scales of Operation

CTF can be practised on practically any scale of farm and level of mechanisation. The only essential criterion is that traffic lanes are separated from the soil used for growing crops. This is generally achieved with relative ease in many bed forming operations but can be more difficult when growing a wide range of crops that have different mechanisation requirements. A key consideration is likely to be the tracked area; the smaller this is the greater the probable cost of implementation, accepting that there is a practical limit on the narrowness of tyres (load/pressure) whose inflation pressure should not exceed 200 kPa. Ideally, pressures should be less than this, so good quality tyres with an enhanced length of contact patch are ideal in helping to avoid problems, for example with repetitive passes in moist soil conditions. Therefore, as with many things, the scale chosen will be a balance between cost and benefit.

Fig. 3.5 12 m wide-span or gantry tractor equipped with a 20-row sugar beet drill. This can be lifted vertically without folding and driven from the field in the machine's longitudinal travel mode

Fig. 3.6 Example of shallow ploughing with a 2 wheel-drive 12 m gantry tractor. Unlike conventional tractors, the machine is not rendered unstable by large offset loads, providing traction is sufficient from the end of the vehicle closest to the draught load

Achieving CTF with a Wide-Span or Gantry Tractor

It has been seen in the preceding sections that a complete system of CTF introduces many challenges, such as the general preclusion of ploughing and lack of flexibility in machine design. However, many of these shortcomings could be avoided if a wide-span or gantry tractor were adopted as the prime mover. Figure 3.5 provides an example of such a machine, which despite being 12 m wide in its field mode, can be driven lengthways on the road, thus precluding the need to fold the attached 20-unit sugar beet drill for example. This type of vehicle is also capable of ploughing (Fig. 3.6) without compromising the wide non-trafficked beds that can represent

Fig. 3.7 The 9.6 m span gantry commissioned and used by a Danish farmer showing it as an onion lifter and full-width cultivator

around 95% of the field area. With such a low proportion of field areas being trafficked, it is conceivable that track lanes could be designed and managed to ensure access, even when soils are at a high moisture content. A further reduction in trafficked area might be achieved with a rail system, as first proposed by Halkett (1858).

These advantages, together with the machine's inherent stability, close control of interlinked implements and potential for staggering rows, makes this machine concept a logical CTF solution for annual horticulture. Specialised versions of these types of machines have been used in the horticultural sector as harvesting aids (Hood et al. 1987; Holt and Tillett 1989; Tillett et al. 1988), often semi-automated and without an onboard driver, as evidenced in the most recent development in the form of NEXAT (see https://www.nexat.de/en/the-system/).

More recent work by Pedersen (2014) reported on a 9.6 m span gantry designed for a Danish farmer and used primarily as an onion elevator, but also capable of full width cultivation. The latter was achieved by changing the centre section of the machine, as illustrated in Fig. 3.7. Pedersen (2014) estimated that the wide span configuration could increase crop yields by up to 20% (primarily on the basis of 6% tracked area compared with 21% for traditional CTF systems), while simulations of five vegetable farms suggested that gross margins would be increased by 44% if a complete wide span system were introduced. In addition, soil structure would be enhanced across a wider area than can be achieved with tractor-based systems and would lead to a 31% reduction in greenhouse gas emissions, largely because of lower nitrous oxide emissions from the soil (Vermeulen and Mosquera 2009). Pedersen (2014) reported that 90% of 31 growers from six different countries who had shown interest in CTF said they would be interested in adopting the wide span technology, providing the machine was robust and economically viable (Tillett and Audsley, 1987).

In terms of field efficiency for harvesting operations (the equivalent researched by Bochtis et al. (2010) and discussed earlier), headland turns would be simplified (all consisting of pirouettes) compared with conventional practice, but in-field travel would still be greater.

Alternatives to CTF

There is currently a burgeoning of autonomous systems with robots of all sizes and designs being researched and introduced for an increasing range of activities (Gil et al. 2023). Many argue that the relative lightness of these machines precludes the need for CTF. McPhee et al. (2020) considered this argument by modelling the maximum loads that might be tolerated without exceeding a critical soil bulk density, whose upper threshold was between 1.3 and 1.4 Mg m^{-3}. Wheel loads associated with grain and potato harvesters were modelled for two different tyre sizes in each case and results indicated that only wheel loads of 0.6 Mg or less avoided exceeding the critical bulk density in the topsoil. The authors considered that remedial tillage would be needed following wheel loads any greater than this.

It is also a fact that many small autonomous machines would be needed to achieve economically and practically viable overall harvest work-rates and this high traffic density poses a further threat to the health of soils, as revealed by Botta et al. (2006). Their research imposed repeated wheel loads of just 1050 kg on a silty clay loam soil that had been in no-till for six years under a wheat/soybean rotation. Not only did the modest pressure of 68 kPa under the cross-ply tyres increase the already high penetration resistance and bulk density (1269 kPa and 1203 Mg m^{-3} respectively) in the top 150 mm of soil, but ten or more passes had an effect to 600 mm depth. Although this scale of repetition may only occur over a longer time period (and radial rather than cross-ply tyres might be used), the topsoil impacts were more immediate and would certainly require remedial tillage within a cultivated rather than a no-till environment as also found by Calleja-Huerta et al. (2023).

CTF Uptake in Horticultural Production Systems

Unfortunately, there is a dearth of information about CTF uptake in the horticultural industry. In Australia, for example, where uptake in the arable sector is buoyant and rapidly expanding (30% of the cropped area in 2016 (GRDC 2016)), the number of horticultural enterprises operating with a CTF system is in single figures (McPhee, personal communication, 2021). Of the 8–10 farms estimated, some are in a seasonal system, often prompted by reliance on contract harvesters for carrots, onions, potatoes, peas and beans (*Phaseolus vulgaris*).

In the Netherlands, CTF within horticultural systems is almost exclusively practised on organic farms and is primarily seasonal in nature, i.e. sCTF. An estimate (Bernaerts, personal communication, 2022) suggests that up to 400 farms (20 of which are non-organic) are using elements of CTF, but only about 10 are close to full CTF implementation.

In the UK, as with Australia, many farmers grow vegetables in beds, but few maintain these from one season to the next. There are two known exceptions, one growing onions and potatoes, the other sweetcorn, asparagus (*Asparagus officinalis*), broccoli, courgettes (*Cucurbita pepo*) and green beans (the latter is Case Study 3.1 in this chapter). The onion grower spent many years operating a sCTF system because the onion harvester was not compatible with the permanent wheel tracks used for other operations. This changed in 2017 when they commissioned a purpose-built harvester that has the elevator's retractable supporting wheel positioned in the next wheel track rather than in the middle of the adjacent bed. This means that when they transition to cereal crops, with implements operating at three times the onion harvester width, they can now deep loosen only the wheel tracks that come within the wider beds. Grain harvesting operates at twice the implement width.

3.5 CTF Case Studies

To assist the reader in understanding how CTF is put into practice, the conversion process for two vegetable farms of contrasting size are used here as examples. Between them they highlight the reasons for conversion and the benefits they have realised since adoption of CTF.

Case Study 3.1: Barfoots of Botley, Hampshire, UK

Farm size: 3500 ha
Soil types: clays to sands and sandy loams
UK crops: sweetcorn, tender-stem broccoli, courgette, squash (*Cucurbita*) (pumpkins), green beans and post CTF, green manures to raise organic carbon levels.
Cultivation system: Before CTF, ploughing, subsoiling and power harrowing. After CTF, tine and disc tillage, power harrowing and targeted subsoiling.

The 3500 ha of Barfoot farms in the UK are complemented by many thousands of hectares in Spain, Senegal and Peru, enabling daily supply to markets around the world. Transition to CTF on their UK farms started in 2014 and was driven by their realisation that to reduce tillage inputs, increase cover cropping and generally improve production efficiency they needed to manage and reduce their high level of soil compaction. Barfoots' philosophy is "never to put a wheel where a seed or plant would ever go" and to reach this goal, farm staff were initiated in the technicalities of CTF, including the importance of high-quality low-pressure tyres.

Transition to CTF

As a baseline for reducing the trafficked area, two of the farms undertook a review of their existing equipment, including track gauges (distance between wheels on an axle), implement widths and tyre sizes. These data were used to establish the wheel track area associated with a rotation of sweetcorn, pumpkins, tender-stem broccoli

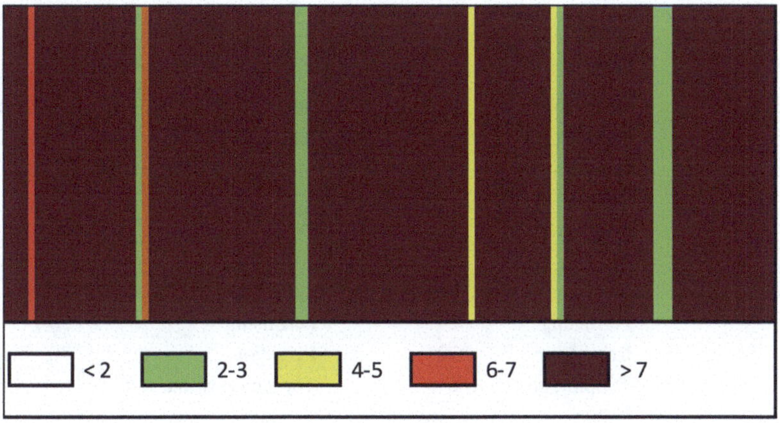

Fig. 3.8 Intensity of tracking in a four-year rotation—number of coincident passes for all crops prior to CTF

(TSB) and beans. This revealed that 100% of the area was trafficked at least once. Figure 3.8 shows that much of the area was trafficked more than seven times, and this was augmented by the extensive foot traffic associated with TSB and an indeterminate amount of random machinery trafficking. Subsoiling, which was carried out at 45 degrees to all the other operations, accounted for an additional 39–50% of coverage depending on the width of the subsoiler.

Having studied their machinery complement, the most logical base width for their CTF system was 5 m with a vehicle track gauge of 1.67 m. Table 3.2 shows that it was possible to fit all the machines into the new system, other than the 4 m wide bunker harvester for sweetcorn and the 1.67 m offset trailed Oxbo harvester for beans. Without these two machines, trafficked area would be reduced to just 37% (Fig. 3.9) from a nominal 7 coincident passes with traditional practice, but where they are employed (albeit on a relatively small proportion of the farm), trafficked area was only reduced to 70%. One of the farms took a progressive approach to conversion, the other made a more immediate transition.

Transitional Soil Conditions

Soon after the farm had converted some of its fields to CTF, the soil within a sweetcorn crop was inspected. This revealed that there was still a distinct compact layer at approximately 130–300 mm in the soil profile. This was apparent in both the non-trafficked bed as well as in the soil immediately adjacent to the wheelings (Fig. 3.10). As a result, there was no detectable difference in crop height, total cob yield or marketable cobs from rows next to wheelings compared with those in the centre of the bed. No comparison between the beds and conventional practice was available as the whole field was managed in the same way. These lack of contrasts

Table 3.2 Operations before and after the introduction of CTF in a rotation of sweetcorn, squash (pumpkins), tender-stem broccoli and beans

Crop area with CTF	Operations	Operating width, m	Operations	Operating width, m
	Before CTF		**After CTF**	
All	Plough + tine cultivation	3	Multi-purpose cultivator	5
	Power harrow	5	Power harrow	5
	Rolling	6		
	Spraying	24	Spraying	25
	Fertilising	24	Fertilising	25
	Irrigation	72	Irrigation	70
	Topping post-harvest	6	Topping post-harvest	5
	Subsoiler	4	Targeted de-compaction	5
	Subsoiler	3		
Sweetcorn	Drilling + plastic layer	4.5	Drilling + plastic layer	5
650 ha	Plastic removal	4.5	Plastic removal	5
	Hoeing	4.5	Hoeing	5
	Hand harvesting to rigs	10	Hand harvesting to rigs[a]	10
	Self-propelled harvester	4.5	Self-propelled harvester	5
	Bunker harvester	4	Bunker harvester	4
	Offloading SP to trailer	2.5		
Tender-stem broccoli 100 ha	Planting	5	Planting	5
	Fleece removal	15	Fleece removal	15
	Hoeing	1.67	Semi-robotic hoeing	5
	Hand harvesting to rigs	15	Hand harvesting to rigs	15
Pumpkins	Drilling	4.5	Drilling	5
75 ha	Hoeing	4.5	Hoeing	5
	Harvesting	2	Hand harvesting into boxes	25[b]
Beans	Drilling	5	Drill + roll	5
67 ha	Hoeing	1.67	Hoeing	5
	Trailed harvester	1.67	Trailed harvester	1.67
Cover crops			Drilled with multi-purpose cultivator	5
			Rolling	15

[a]Rigs are various elements associated with the harvesting process such as elevators and towed packing units

[b]Before CTF, pumpkins were hand harvested into rows from an indeterminate width, whereas with the CTF system, there is a clear division every 25 m

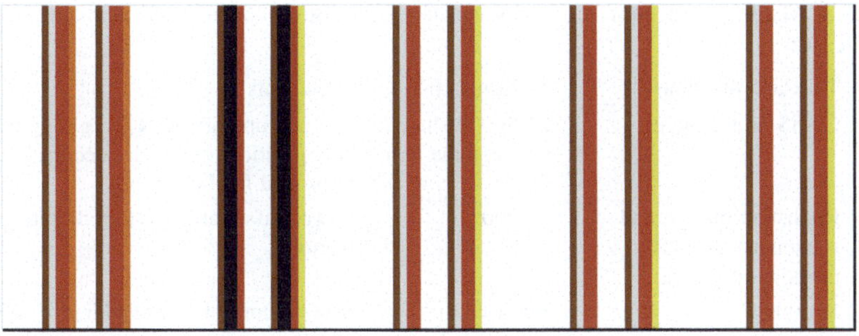

Fig. 3.9 Tracking for all crops without the 4 m wide bunker harvester or trailed offset 1.67 m Oxbo harvester. Tracked area is just under 37%. Colours relate to the legend in Fig. 3.8 with the addition of black which are those used for chemical applications

Fig. 3.10 Residual compact layers between 130 and 300 mm depth (**a** and **b**) and in an extracted block of soil (**c**) in fields recently converted to CTF

reflect the time needed for soil to recover from compaction naturally and the advantage that might be gained from targeted deep loosening.

The Cost of Introducing CTF

Having made a progressive rather than a rapid change to CTF (over four years), the farm limited expenditure to a large degree, with the marginal cost being the extra investment required to ensure that replacement machines fitted the CTF system.

Table 3.3 Marginal (additional) costs associated with conversion to CTF across two of the Barfoots' farms

Year	Task and machine	Marginal cost, £	Comments
2013	GNSS & auto-steer	£100,000 shared between the two farms	Planned upgrade to RTK, giving repeatable positioning, greater accuracy and improved field efficiency
2013	Replacing plough and subsoiler with multi-purpose cultivator	15,000	Fewer passes and less fuel despite bigger tractor
2013	Spraying	Minimal to £600 + 1 day labour	Section control and blanking of some nozzles and boom extension
2015	Sweetcorn drill and poly layer replaced	None	Natural progression, saved 1 tractor and 4 people by virtue of new technology
2013	Sweetcorn hoeing	1 day labour	Moved hoeing elements to fit
2013	Specialist sweetcorn harvester	3 day's labour	Altered the header
2016	Replaced 4 m subsoiler with 5 m subsoiler	Natural progression	Lower cost per unit area worked, assisted by non-trafficked soil
2016	TSB hoe replaced with inter-row precision cultivating weeder	£50,000	Total cost £107,000, the marginal cost was to gain extra 0.5 m width to fit CTF system, which also covers more ground. Reduces need for hand weeding.
2017	Rolling, 6 m rolls	None attributed	Existing rolls needed replacement and natural progression was to 15 m

Table 3.3 provides an overview of these costs across two farms. Several machines were either already the required width or could be adjusted without modification. The latter included 6 m toppers that were simply used with an overlap, while fertiliser spreaders had an adjustment to increase their spreading width to 25 m. Each farm needed its own GNSS upgrade and about 50% of the costs associated with a base station and ten receivers were shared between the two farms, resulting in total costs of £58 k on one farm and £59 k on the other.

Set against these costs were savings in labour but these were difficult to quantify because Barfoots had taken on more land during the period. In terms of fuel, there was an estimated 10–15% reduction per hectare while power demand dropped by 25–30% from around 4.6 kW/ha to 3.4 kW/ha.

Notes on the Barfoots Case Study

- Traditional practice was resulting in excessive compaction marked by 100% coverage and by up to seven passes over approximately 95% of the soil. The 5 m CTF system reduced tracking from a nominal 7 passes over the whole area to just 70% of the area (including the two incompatible machines) and to 37% where all the machines are aligned.

- Removing ploughing and some tillage operations has reduced the power demand on the farms by between 25%–30% from around 4.6 $kWha^{-1}$ to 3.4 $kWha^{-1}$.
- The large number of machines associated with sweetcorn means that it is a crop that benefits highly from CTF.
- Introducing more appropriate tyres and/or tracks could reduce the trafficked area further, as could increasing the base width of the machines, but the latter would entail a major new phase in the transition.
- Avoiding compaction has allowed greater use of minimum tillage to create a seedbed but deeper compaction will require further time or the use of targeted deeper cultivations and/or cover crops to ameliorate.
- Minimum tillage helps to conserve moisture in the seedbed, which is crucial in times of warmer weather and water abstraction restrictions. The absence of ploughing has caused challenges with the seed drill, which they are planning to replace with a drill that can deal with more surface residues.
- Return on investment in CTF has been difficult to quantify but savings in fuel, labour and power have been identified.
- Rented land meant that CTF could not be used universally because no other farmers in the area were on the same system.

Case Study 3.2: Duco and Mart Verbrugge, CTF Organic Horticulture Growers, Hengstdijk, Netherlands

Duco Vebrugge farms clay soils on polder land (land reclaimed from the sea) in partnership with his father Mart Verbrugge.

Farm Details

- **Farm size:** 40 ha
- **Soil type:** deep clay—varying from lighter to heavier clay
- **Crops:** red beet (*Beta vulgaris*), potatoes, seed onions, alfalfa (*Medicago sativa*), red cabbage, squash, celery (*Apium graveolens*) and carrots
- **Cultivation system:** reduced tillage; occasional ploughing for weed control—power harrow only used for small-seeded crops.

The farm converted to CTF in 2008, modifying two tractors (67 and 112 kW from a track gauge of 1.8 m to 3.2 m (Fig. 3.11) and installing RTK GNSS. They used advisers and other farmers to help with the conversion, adopting the best aspects of other growers' production systems:

"We have used other farmers' good results and are using other farmers' systems!"

All tractors, implements and machinery are on a 3.2 m track gauge apart from lucerne and carrot harvesters; cultivators and harrows operate across 3 m beds. This represented a significant investment and initially there were problems with the technology, but these have largely been resolved.

Fig. 3.11 Duco Verbrugge with one of the tractors whose axles were extended to give a track gauge of 3.2 m. Note the reinforcement of the front axle and the "cotton reel" extensions to the rear axle

The greatest challenge was modifying the potato harvester (for which they used a 30% innovation grant subsidy) that involved fitting a track under the machine to reduce the load on the axle.

Experience of CTF

Duco and Mart are convinced that their CTF system provides better soil structure (Fig. 3.12) and more even crop quality than a traditional cultivation system. This helps them meet supermarket quality requirements, for example in terms of the size and shape of carrots. They plough sparingly and only when needed for weed control, but they are considering the occasional use of a shallower plough on the wheelings where weed pressure tends to be higher.

The use of poultry manure, mushroom compost and sugar beet "vinasse" (a by-product of alcohol production from sugar beet) is an important part of their system to maintain soil organic matter and to supply nutrients.

Duco and Mart thought that crop yields may have increased slightly ("not spectacularly") but noted that soil was easier to cultivate and that they could travel on the land sooner after rain (soil drains better), improving access and giving an earlier start to the season. The CTF system also made interrow weeding easier.

Fig. 3.12 Duco and Mart contrast soil from the non-trafficked bed (left) with that from beneath the wheelings (right)

Duco and Mart were considering the use of smaller tractors to reduce compaction, especially when ploughing. They noted that "the ground is less strong after ploughing".

3.6 Summary/Conclusions

Compaction is an increasingly insidious part of vegetable production systems that is compromising the health and function of soils. The intense and randomly applied stresses created by heavier and heavier machines increases the strength of soils, reduces their porosity and their ability to drain efficiently. This increases the cost of tillage and exacerbates run-off and soil loss. Controlled traffic farming (CTF) addresses these issues by confining all traffic to the least possible area within the constraints of cost and practicality. Thus, farms that might have been creating repeated tracking over 100% of their land can reduce this to less than 40% with an associated 25–30% reduction in power required for tillage.

Introducing CTF requires the alignment of operating widths and track gauges of all field machines to minimise tracked areas. This requires a whole farm approach, attention to detail and a commitment to continual improvement over a number of years. Gradual transition is possible using a seasonal system that precludes harvest operations but may only have limited benefits. Changes in agronomy are needed to address lower tillage inputs and more amenable soil conditions.

CTF can be achieved on any scale of farm with either conventional tractor-based systems or by gantry tractors. The latter currently make for a more expensive transition but reduce tracked areas and are more efficient in many aspects of their operation. Uptake of CTF in the arable sector has been buoyant, particularly in Australia but has been limited for horticultural operations reflecting the more complex processes involved. However, the benefits of conversion are likely to be much greater than in the arable sector, particularly as they relate to energy use and the quality and consistency of products.

Competing Interests The authors have no conflicts of interest to declare that are relevant to the content of this chapter.

References

Antille DL, Peets S, Galambošová J, Botta GF, Rataj V, Macak M, Tullberg JN, Chamen WCT, White DR, Misiewicz PA, Hargreaves PR, Bienvenido JF, Godwin RJ (2019a) Review: soil compaction and controlled traffic farming in arable and grass cropping systems. Agron Res 17(3):653–682. https://doi.org/10.15159/AR.19.133

Antille DL, Chamen WC et al (2019b) Controlled traffic farming in precision agriculture. In: Stafford J (ed) Precision agriculture for sustainability. Burleigh Dodds Science Publishing, Cambridge

Bochtis DD, Sørensen CG, Green O, Moshu D, Olesen J (2010) Effect of controlled traffic on field efficiency. Biosyst Eng 106:14–25

Botta GF, Jorajuria D, Rosatto H, Ferrero C (2006) Light tractor traffic frequency on soil compaction in the Rolling Pampa region of Argentina. Soil & Tillage Research, 86:9–14

Calleja-Huerta A, Lamandé M, Green O, Munkholm LJ (2023) Vertical and horizontal stresses from lightweight autonomous field robot during repeated wheeling. Soil Tillage Res 233: 105790

Carballido J, Perez-Ruiz M, Emmi L, Aguera J (2014) Comparison of positional accuracy between RTK and RTX GNSS based on the autonomous agricultural vehicles under field conditions. Appl Eng Agric 30(3):361–366. https://doi.org/10.13031/aea.30.10342

Chamen WCT (2011) The effects of low and controlled traffic systems on soil physical properties, yields and the profitability of cereal crops on a range of soil types. PhD thesis, Cranfield University, Cranfield

Chamen WCT, Audsley E (1993) A study of the comparative economics of conventional and zero traffic systems for arable crops. Soil Tillage Res 25:369–390

Chamen WCT, Longstaff DJ (1995) Traffic and tillage effects on soil conditions and crop growth on a swelling clay soil. Soil Use Manag 11:168–176

Chamen WCT, Chittey ET, Leede PR, Goss MJ, Howse KR (1990) The effect of Tyre-soil contact pressure and zero traffic on soil and crop responses when growing winter wheat. J Agric Eng Res 47(1):1–21

Chamen WCT, Watts CW, Leede PR, Longstaff DJ (1992) Assessment of a wide span vehicle (gantry) and soil and cereal crop responses to its use in a zero traffic regime. Soil Tillage Res 24: 359–380

Chamen WCT, McPhee JE, Pedersen HH, Carter LM (2025) Chapter 13: developments in controlled traffic farming (CTF) in precision agriculture. In: Stafford J (ed) Precision agriculture for sustainability, 2nd edn. Burleigh Dodds Science Publishing, Cambridge

da Luz FB, Gonzaga LC, Castioni GAF et al (2023) Controlled traffic farming maintains soil physical functionality in sugarcane fields. Geoderma 432:116427

Derpsch R, Friedrich T, Kassam A, Hongwen L (2010) Current status of adoption of no-till farming in the world and some of its main benefits. Int J Agric Biol Eng 3(1):1–25

Dickson JW, Ritchie RM (1996) Zero and reduced ground pressure traffic systems in an arable rotation. 1. Cultivation power requirement. Soil Tillage Res 38(1–2):71–88

Dickson JW, Campbell DJ, Ritchie RM (1992) Zero and conventional traffic systems for potatoes in Scotland, 1987-1989. Soil Tillage Res 24(4):397–419

Domzal H, Glinski J, Lipiec J (1991) Soil compaction research in Poland. Soil Tillage Res 19(2–3): 99–109. https://doi.org/10.1016/0167-1987(91)90079-d

Esteban DAA, de Souza ZM, Tormena CA, Lovera LH, de Lima E, Souza Lima E et al (2019) Soil compaction, root system and productivity of sugarcane under different row spacing and controlled traffic at harvest. Soil Tillage Res 187:60–71

Gil G, Casagrande D, Pérez Cortés L, Verschae R (2023) Why the low adoption of robotics in the farms? Challenges for the establishment of commercial agricultural robots. Smart Agric Technol 3:100069

Godwin RJ, White DR, Dickin T, Kaczorowska-Dolowy M, Millington WAJ, Pope EK et al (2022) The effects of traffic management systems on the yield and economics of crops grown in deep, shallow and zero tilled sandy loam soil over eight years. Soil Tillage Res 223:105465

GRDC (2016) Farm practices survey report 2016. Grains Research & Development Corporation, PO Box 5367, Kingston, ACT 2604

Håkansson I, Reeder RC (1994) Subsoil compaction by vehicles with high axle load extent, persistence and crop response. Soil Tillage Res 29:277–304

Halkett PA (1858) On guideway agriculture: being a system enabling all the operations of the farm to be performed by steam-power. J Soci Arts 7(316):41–58. https://doi.org/10.2307/41334925

Harrison DJ, Dawson EM, Birkenshaw JE, Davies ACW (1985) The effect of soil compaction on the yield and quality of vegetable crops. Res Dev Agric 2(2):71–75

Hefner M, Labouriau R, Nørremark M, Lakkenborg Kristensen H (2019) Controlled traffic farming increased crop yield, root growth and nitrogen supply at two organic vegetable farms. Soil Tillage Res 191:117–130

Holt JB, Tillett ND (1989) The development of a nine metre span gantry for mechanised production and harvesting of cauliflowers and other field vegetables. J Agric Eng Res 43:125–135

Hood CE, Williamson RE, Garrett TR, Young RE (1987) Straddle vehicles for horticultural crop production – a review. Acta Hortic 210:35–42

Hussein MA, Antille DL, Kodur S, Chen GN, Tullberg JN (2021) Controlled traffic farming delivers improved agronomic performance of wheat as a result of enhanced rainfall and fertiliser nitrogen use efficiency. Acta Agric Scand B Soil Plant Sci 71(5):377–398

Isbister B, Blackwell P, Reithmuller G, Davies S, Whitlock A, Neale T (2013) Controlled traffic farming technical manual. Department of Agriculture and Food, Perth

Botta GF, Jorajuria D, Rosatto H, Ferrero C (2006) Light tractor traffic frequency on soil compaction in the rolling Pampa region of Argentina. Soil Tillage Res 86:9–14

Kaczorowska-Dolowy M, Godwin RJ, Dickin E, White DR, Misiewicz PA (2019) Controlled traffic farming delivers better crop yield of winter bean as a result of improved root development. Agron Res 17(3):725–740

Keller T, Or D (2022) Farm vehicles approaching the weights of sauropods exceed safe mechanical limits for soil functioning. PNAS 119(21). https://doi.org/10.1073/pnas.2117699119

Keller T, Sandin M, Colombi T, Horn R, Or D (2019) Historical increase in agricultural machinery weights enhanced soil stress levels and adversely affected soil functioning. Soil Tillage Res 194: 1–12. https://doi.org/10.1016/j.still.2019.104293

Kingwell R, Fuchsbichler A (2011) The whole-farm benefits of controlled traffic farming: an Australian appraisal. Agric Syst 104(7):513–521. https://doi.org/10.1016/j.agsy.2011.04.001
Kroulik M, Kumhála F, Hula J, Ohnzik I (2009) The evaluation of agricultural machines field trafficking intensity for different soil tillage technologies. Soil Tillage Res 105(1):171–175
Kuipers H, Zande JC et al (1994) Quantification of traffic systems in crop production. In: Soane BD, Ouwerkerk CV (eds) Soil compaction in crop production. Elsevier Science B.V, pp 417–445
Lamers JG, Perdok UD, Lumkes LM, Klooster JJ (1986) Controlled traffic farming systems in The Netherlands. Soil Tillage Res 8:65–76
Lanyon DM, Cass A, Hansen D (2004) The effect of soil properties on vine performance: a review. Am J Enol Vitic 36:125–131
McGarry D (2003) Tillage and soil compaction. In: Torres LG (ed) Conservation agriculture: environment, farmers experiences, innovations, socio-economy, policy. Netherlands Kluwer Academic Publishers, pp 307–316
McHugh AD, Tullberg JN, Freebairn DM (2009) Controlled traffic farming restores soil structure. Soil Tillage Res 104(1):164–172
McPhee JE (2020) Controlled traffic for vegetable production: challenges, benefits and opportunities. Submission in fulfilment of the requirements for the degree of doctor of philosophy. Tasmanian Institute of Agriculture, University of Tasmania
McPhee JE, Aird PL (2013) Controlled traffic for vegetable production: part 1. Machinery challenges and options in a diversified vegetable industry. Biosyst Eng 116(2):144–154. https://doi.org/10.1016/j.biosystemseng.2013.06.001
McPhee JE, Pedersen HH (2017) Economic modelling of controlled traffic for vegetable production based on the use of wide span tractors. Aust Farm Bus Manag J 14:71–88
McPhee JE, Braunack MV, Garside AL, Reid DJ, Hilton DJ (1995a) Controlled traffic for irrigated double cropping in a semi-arid tropical environment: part 2, tillage operations and energy use. J Agric Eng Res 60(3):183–189
McPhee JE, Braunack MV, Garside AL, Reid DJ, Hilton DJ (1995b) Controlled traffic for irrigated double cropping in a semi-arid tropical environment: part 3, timeliness and trafficability. J Agric Eng Res 60(3):191–199. https://doi.org/10.1006/jaer.1995.1013
McPhee JE, Aird PL, Hardie MA, Corkrey SR (2015) The effect of controlled traffic on soil physical properties and tillage requirements for vegetable production. Soil Tillage Res 149:33–45. https://doi.org/10.1016/j.still.2014.12.018
McPhee JE, Antille DL, Tullberg JN, Doyle RB, Boersma M (2020) Managing soil compaction – a choice of low-mass autonomous vehicles or controlled traffic? Biosyst Eng 195:227–241
Misiewicz PA, Galambošová J (2023) Controlled traffic farming. In: Zhang Q (ed) Encyclopedia of digital agricultural technologies. Springer, Cham. https://doi.org/10.1007/978-3-031-24861-0_105
Nawaz MM, Noor MA, Latifmanesh H, Wang X, Ma W, Zhang W (2023) Field traffic-induced soil compaction under moderate machine-field conditions affects soil properties and maize yield on sandy loam soil. Front Plant Sci 14:1002943
Neale TJ, Tullberg JN (1996) Controlled traffic in lucerne hay production. Paper presented at the conference on engineering in agriculture and food processing, University of Queensland, Gatton College, Lawes, Queensland, Australia November 24–27, p 97
Neilsen S (2008) Runoff and constituent loss at Moonggoo – 1999-2008 summary. Cropping Central – Central Queensland sustainable farming systems project 41, p 10
Pedersen HH (2014) Wide span technology for efficient and environmental friendly vegetable production. PhD thesis, Aarhus University, Denmark. https://www.researchgate.net/publication/365174711_Wide_Span_technology_for_efficient_andenvoronmental_friendly_vegetabel_production
Pedersen HH (2015) Evaluation of onion production in sandy soils by use of reduced tillage and controlled traffic farming with wide span tractors. Acta Technol Agric 18(3):74–82

Powrie J, Bloomer D (2010) The dirt on controlled traffic potatoes. Retrieved November 5, 2018, from http://www.landwise.org.nz/projects/the-dirt-on-controlled-traffic-potatoes/
Rodgers D, McPhee J, Aird P, Corkrey R (2018) Soil arthropod responses to controlled traffic in vegetable production. Soil Tillage Res 180:154–163
Shaheb MR, Venkatesh R, Shearer SA (2021) A review on the effect of soil compaction and its management for sustainable crop production. J Biosyst Eng 46:417–439
Stirling GR (2008) The impact of farming systems on soil biology and soilborne diseases: examples from the Australian sugar and vegetable industries – the case for better integration of sugarcane and vegetable production and implications for future research. Australas Plant Pathol 37(1): 1–18
Tamirat TW, Pedersen SM, Farquharson RJ, de Bruin S, Forristal PD, Sørensen CG et al (2022) Controlled traffic farming and field traffic management: perceptions of farmers groups of northern and Western European countries. Soil Tillage Res 217:105288. https://doi.org/10.1016/j.still.2021.105288
Tillett ND, Audsley E (1987) The potential economic benefits of gantries for leaf vegetable production. J Agric Eng Res 36(1):31–44
Tillett ND, Holt JB, Chestney AAW, Reed JN (1988) Experimental field gantry for leaf vegetable production: specification, design and evaluation. J Agric Eng Res 41(1):53–64
Tullberg JN (2000) Wheel traffic effects on tillage draught. J Agric Eng Res 75(4):375–382. https://doi.org/10.1006/jaer.1999.0516
Tullberg J, Antille DL, Bluett C, Eberhard J, Scheer C (2018) Controlled traffic farming effects on soil emissions of nitrous oxide and methane. Soil Tillage Res 176:18–25. https://doi.org/10.1016/j.still.2017.09.014
van Balen D, Cuperus F, Haagsma W, de Haan J, van den Berg W, Sukkel W (2023) Crop yield response to long-term reduced tillage in a conventional and organic farming system on a sandy loam soil. Soil Till Res 225:105553
Vermeulen GD, Mosquera J (2009) Soil, crop and emission responses to seasonal-controlled traffic in organic vegetable farming on loam soil. Soil Tillage Res 102(1):126–134
Vermeulen G, Tullberg J, Chamen W (2010) Controlled traffic farming. In: Dedousis AP, Bartzanas T (eds) Soil engineering, Part 2, vol 20. Berlin Heidelberg, Springer, pp 101–120
Yigit CO, Bezcioglu M, Ilci V, Ozulu IM, Alkan RM, Dindar AA et al (2022) Assessment of real-time PPP with Trimble RTX correction service for real-time dynamic displacement monitoring based on high-rate GNSS observations. Measurement 201:111704

Chapter 4
Precision Integrated Pest Management in Horticulture; Pest and Disease Forecast, Detection, and Management

Mark W. Ramsden, Tito Caffi, and Margherita Furiosi

Abstract Demand for large quantities of horticultural produce, at a low price, is driving growers towards more intensive production systems. This in turn drives application of digital technologies to improve targeted management of crops, and the associated risks posed by a range of invertebrates and pathogens. Predictions about the timing and severity of these pests, along with advances in landscape and crop monitoring, support improved decision making, and help avoid unnecessary applications of pesticides. In this chapter, we summarise the current options available, and future opportunities for managing horticultural pests using precision technology, within the context of a transition towards wider uptake of more holistic approaches to crop protection.

Keywords Integrated pest management · Decisions support systems · Monitoring · Precision horticulture

4.1 Introduction

There is increasing demand for horticultural crops, driving intensively managed production systems. Low tolerance in the market for the presence of pathogens, invertebrates (either pests or beneficial species) or associated damage, alongside low tolerance for chemical use, is putting these production systems under pressure (Pimentel 2005; Tilman et al. 2011; Kunjwal and Srivastava 2018; Nitzko et al. 2022). Integrated pest management (IPM) aims to protect crops against pests (i.e., any species, strain, or biotype of plant, animal, or pathogen injurious to plants or plant products; Bajwa and Kogan 2002) while keeping pesticides and other

M. W. Ramsden (✉)
ADAS, Cambridge, UK
e-mail: Mark.Ramsden@adas.co.uk

T. Caffi · M. Furiosi
USCS, Milan, Italy
e-mail: Tito.Caffi@unicatt.it; Margherita.furiosi@unicatt.it

P. Newell Price et al. (eds.), *Precision Agriculture in Field Horticulture*, Progress in Precision Agriculture, https://doi.org/10.1007/978-3-031-96534-0_4

interventions to levels that are economically and ecologically justified, thus minimizing risks to humans and the environment (Barzman et al. 2015). The eight principles of IPM, as laid out in the EU's Sustainable Use Directive (SUD 2009/128/EC) define the key aspects that farmers, advisors, and researchers should consider in their approach to crop protection. These principles provide some guidance by which horticultural crops can be managed holistically to prevent, tolerate, predict, mitigate, and treat pest infestations. The principles addressing prevention and suppression, and monitoring and decision making, underpin low-input strategies, and ensure that treatment is only applied where necessary. This relies on combining preventative tactics into an overall strategy, and using observation networks, pest forecasts, diagnostic tools, and thresholds to make informed decisions about management options. Good agronomic planning and practice at establishment, such as crop rotation, variety selection, planting and sowing techniques are fundamental. Cultural practices, such as prompt removal of plant residue from previous crops can, for example, greatly reduce the risk of carryover of infestations to following crops. Pesticide resistance is a strong driver of increasing uptake of IPM strategies in horticulture, as it has reduced the number of possible chemical management strategies available.

Transition towards more holistic, integrated approaches to pest management is necessary, but challenging (Fraser et al. 2011; Barzman et al. 2015). Integrated approaches are more complicated to manage, and many tactics involve trade-offs between the management of different pests in the crop, or on the yield or quality of the final product. Management of horticultural crops is becoming ever more targeted to achieve sufficient prevention and/or suppression of infestations, while minimising cost inputs. In this chapter, we look at how implementation of precision horticulture fits into wider holistic IPM strategies, though more detailed forecasting, and detection and management of pests in horticultural crops. We then discuss the barriers and incentives to the uptake of decision support tools for precision management of pests, and conclude with the future needs and opportunities for the industry.

Precision IPM in Horticulture: Forecasting, Detecting, and Managing pests in Horticultural Crops

Secure food supply for future generations requires environmentally safe and sustainable production. Different types of pests and abiotic stress factors (e.g. deficiency of nutrients and drought) reduce quantity and quality of agricultural and horticultural crops worldwide. A main goal of modern agriculture is to reduce intensive fertilizer and pesticide use, decrease heavy exploitation of natural resources (water, soil, energy), and arrest or ideally reverse the decline in biodiversity while maintaining or enhancing crop yield/quality (Rossi et al. 2012). Integration of digital innovations and technologies with traditional agricultural practices can contribute towards sustainable and secure crop/food production (Shepherd et al. 2018), and aligns with the aims and spirits of Sustainable Development Goals adopted by all United Nations Member States in 2015 (FAO 2024). Precision horticulture is defined here as management of spatial and temporal variability in fields using information and communications technologies (ICT) (Blackmore et al. 2003), or more simply doing "the right thing, at the right time in the right place" (Gebbers and Adamchuk 2010). This is opposed to more conventional practices where the various crop treatments, such as irrigation, application of fertilizers, pesticides and herbicides are uniformly applied to the entire field, without accounting for any variability within the field. Precision horticulture aims to improve target related technologies, environmental effects, economic outcomes, adoption rates and drivers (Pierpaoli et al. 2013), using real-time, robust mapping systems for crop, soil, and environment variables to produce guidance for management decisions (Mahlein 2016).

There are three aspects of IPM to which the principles of precision horticulture can be applied. Firstly, tools providing pest forecasts enable temporal targeting of pests, by providing information on which pests may represent risks to crops during vulnerable periods. Secondly, tools that provide data on the presence of pests within the crop, support more targeted management as part of precision monitoring programs. Finally, more holistic approaches to pest management both maximise preventative actions across a range of pests, and balance trade-offs between pests, to ensure actions are targeted appropriately. A holistic strategy would, for example, include actions that promote a diverse and robust population of natural enemies of crop pests in and around the farmed landscape (Chaplin-Kramer et al. 2011). These three routes enable farmers and their advisors to take proportionate action according to needs to ensure economically viable, and environmentally sustainable crops.

4.2 Forecasting pests and Diseases in Horticultural Crops

Weather-based forecast models and observation-based thresholds are commonly used for predicting risk to crops from invertebrates and diseases (Magarey et al. 2002). These are used for scheduling management operations including scouting and, most often, targeted pesticide treatments (Isard et al. 2015). The development of forecasts usually involves four steps; i) identification of relevant environmental inputs (e.g. climate data), ii) model construction and parameterisation, iii) validation and iv) implementation, which usually requires development of an associated Decision Support System (DSS) that translate model outputs into quantifiable levels of risk (Rossi et al. 2012). Pest specific forecast and associated DSS are increasingly being combined into holistic crop protection support systems. Online systems can increasingly automate collation of input data, such as open access climate data, user-provided data (e.g. growers' own weather stations) or additional pay-per-use services. Where online platforms are sufficiently independent and robust, they can also provide a source of acceptable criteria for justifying management decisions, whether to regulatory bodies, wholesalers, food processers or customers (Rossi et al. 2012).

Mathematical and/or statistical models for invertebrate population dynamics and disease epidemiology represent the basis of most forecasting tools applied in horticulture and, broadly, in IPM. A model provides a simplified representation of reality, where the reality in our case is the pest triangle (i.e. pest, crop, and environment), based on the current available knowledge (Rossi et al. 2010). Empiric and mechanistic models are the most relevant categories of models in supporting decision-making in IPM. Empiric models, also called data-based models, organize data, and standardise their relationship in terms of mathematical or statistical representations (e.g., correlation between pest abundance and air temperature). These provide useful insight to explore the relationships within a system that are unknown or poorly known. Mechanistic, or process-based, or fundamental, models describe a process (e.g., pest population dynamics/epidemics) based on the underlying functional mechanisms of the process. Mechanistic models are crucial to evaluate the biological responses as a function of one or more environmental independent variables (e.g., air temperature, relative humidity, etc.) (Rossi et al. 2010, 2019; Caffi et al. 2007).

The decision about the type of modelling approach to implement is crucial to develop the most suitable forecasting system, and is an important consideration for application in precision horticulture. Empirical models have historically been the most common type, however these are often limited as they are not directly linked to the underlying processes of pest population distribution, abundance, and impacts with the environmental driving variables. Mechanistic models can more reliably describe how biotic and abiotic drivers influence an individual pests life-history and, in turn, the overall population dynamics/epidemics. Furthermore, mechanistic models seek the dynamical description of the biological and physiological responses in relation to the population dynamics and epidemics, both in space and time, of the environmental drivers. The bottom-up approach used in mechanistic models requires that the underlying mechanisms of the processes involved (population dynamics/epidemics) are previously known in detail. The amount of data needed for the development of empiric models is high, but their elaboration is relatively simpler compared to the development of mechanistic models, and can provide a more detailed in-depth simulation of the processes involved (Rossi et al. 2012, 2019). Mechanistic models have a further advantage in that they can be more easily updated or expanded as further data becomes available.

Whether they are empiric or mechanical, forecasting systems can provide growers with useful information such as an estimation of first arrival of pests in the crop, or more specific predictions on when management actions may be best targeted at specific life stages (Collier et al. 2020). Forecast systems are best combined with knowledge of crop variety, growth stage and crop condition to give an overall assessment of the likelihood of crop injury. This knowledge is usually achieved through inspection of the crop by the grower or the agronomist. If a DSS predicts a low likelihood of pest activity and/or the crop is not at a vulnerable growth stage, no further action is necessary beyond routine crop monitoring. Where these systems suggest that pest activity is likely, more detailed monitoring of the crop, either through more detailed inspection during crop walks, or collection of targeted remote data, is required to assess pest levels and the potential for economic injury, using thresholds—where these are available (Ramsden et al. 2017) (Fig. 4.1).

Fig. 4.1 Example of grapevine disease. Image credit Tito Caffi

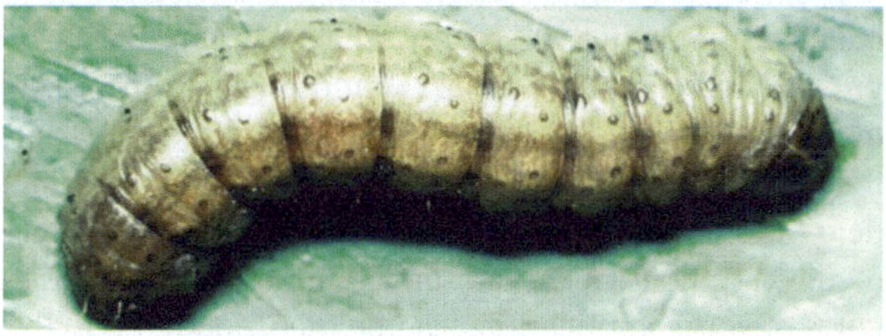

Fig. 4.2 *Agrotis* larva. (Image credit ADAS)

Case Study Example—Precision Timing for Disease Management in Organic Vineyards

Vite.net® was developed for sustainable management of vineyards and is intended for use by the vineyard manager. This holistic DSS consists of two parts: i) an integrated system for real-time monitoring of the field crop components (air, soil, plants, pests, and diseases); and ii) a web-based tool that analyses these data by using advanced modelling techniques and then provides up-to-date information for managing the field crop in the form of alerts and decision supports. The application of this DSS has proved to be effective for the timely control of pests. For instance, vite.net® was tested in 21 organic farms in Italy (which ranged in size from 1 to 180 ha) and allowed, over two seasons, the same level of grape protection obtained with the usual farm practice with an average reduction in the total amount of copper (applied to prevent and treat a range of fungal and bacterial diseases) of 37% because of both reduced doses and fewer applications (Rossi et al. 2014a). This saving was equivalent to 195 €/ha/year for the growers. The same DSS was then applied on a large scale across Italy, confirming the positive results (Rossi et al. 2014b) and is now used by 400 public/private advisers on about 16,000 ha of Italian vineyards (Fig. 4.2).

Case Example—Precision Timing for cutworm Management in Root Crops

Cutworms are larvae of a few species of moth (e.g. *Agrotis* species) that feed at the base and roots of various crops. Eggs are laid in late spring, and the first three instar feed on surface vegetation, before burrowing into roots. Once the larvae (Fig 4.2) have moved into the roots, management is more difficult as they cannot be controlled using contact insecticides. Forecast models that provide guidance on the likely timing and intensity of attacks facilitate more precise timing of management, improving control and reducing the frequency of unnecessary pesticide applications (Bowden et al. 1982, Esbjerg and Segsgaard 2019). As adults continue to lay eggs in the crop after initial arrival, several 'batches' of larvae, at different instars, can be present in the crop at any given time. An empirical model developed and used in the UK (Bowden et al. 1982) provides an estimate of the number of batches of first, second, or third instar larvae likely to be active at any time. When four or more batches are predicted, monitoring and appropriate treatment is recommended to prevent high numbers of larvae moving into the crop roots. The model is run from the first arrival of adults into the crop, based on crop observations, or by default in the UK as the first day after first June where temperature exceeds 12 °C degrees as a default to predict adult moth arrival. Daily temperature is used to determine the rate of development of larvae, and rainfall is used to determine larvae mortality. The dates of spray applications can be incorporated into the model, and are deemed to be 100% effective at removing current cutworm larvae—but the sprays do not prevent subsequent batches (Bowden et al. 1982). This forecast model is hosted on the IPM Decisions platform (IPM Decisions 2022), validated for use in the UK. While the model can be used to identify periods during which a crop is vulnerable to potentially high numbers of 1–3 instar larvae, management of infections still relies on accurate identification of the larval activity in the crop through appropriate monitoring, and subsequent application of an effective intervention. While management remains highly dependent on chemical insecticides, these are limited in the case of cutworm not only to the above ground growth stages, but as applications also need to be applied early in the morning or in the evening as during the day larvae seek refuge in cracks and crevices, making it difficult for spray applications to be applied effectively (Esbjerg and Segsgaard 2019). As part of an IPM program, entomopathogenic nematodes (EPN) offer a potential alternative response to pest infestations, however most candidates for cutworm are active in the soil, and so target later instars (Kumar et al. 2022). In this case, the Bowden model could be adapted to provide guidance on when EPN, rather than insecticides, should be applied to crops.

Application of Forecasts in Precision Horticulture

Despite every effort to develop and validate forecasting models, accurate prediction of risk cannot be guaranteed at all locations (Magarey and Isard 2017). The models are, for example, dependent on the accuracy of weather data at any given location.

Gridded and/or extrapolated weather data can reduce the precision and accuracy of predictions. Forecast models, especially empiric models, make assumptions about the pest and associated ecology that may not apply to every location. Variation in overwinter diapause in cabbage root fly, *Delia radicum,* for example means that generic forecasts that do not account for this fail to effectively predict emergence at all locations (Collier et al. 2021). This natural variation in species ecology makes validation of models in areas outside of their original development crucial for effective uptake. Even when models accurately predict pest ecology, and suitable weather data is available, if the outputs are not readily applicable to a management decision, they are not able to improve pest management. Forecast development therefore requires collaboration of pest experts, crop experts, climate experts, agronomists, social scientists, and software developers to create useful models and associated DSS that improve precision application of IPM in horticulture (Magarey and Isard 2017). A digital road-map for sustainable management of crops should be defined, as was recently proposed for fruit and foliar diseases by Hasanaliyeva and collaborators (Hasanaliyeva et al. 2022).

4.3 Detecting Pests and Diseases in Horticultural Crops

Pest forecasts can provide useful information on predicted risks to crops. For precision horticulture, however, data on actual pest activities and abundance in the field by the grower and/or advisor is essential. Remote sensing can be used to detect some infestations, using technology such as radar to monitor species migration (Leskinen et al. 2011), video equipment to observe insect populations in flight (Preti et al. 2021), and various other automated detection and monitoring technologies (Cardim Ferreira Lima et al. 2020). The use of these innovative technologies has the potential to identify infestations much earlier, and at wider scales than is possible with human observations, and can free up time for collection of more detailed information using other diagnostic tools. Whether automated data collection is used, or a 'hands-on' approach in the field, these observations provide complementary information independent of modelled or remote detected data. The decision maker(s) can then collate all this information and put it into the context of other factors impacting the crop at that time. Detection of pests and diseases using automated systems can provide a more consistent, scalable, and quantifiable measure of key metrics relating to pest/disease activity, and can enable more information to be collected on actual pest activity. Direct observations by the grower/advisor in the field has the advantage of a more holistic assessment at the point of infestation, picking up factors that automated systems may not.

Remote Detection of Diseases and invertebrate Pests

There are several ways in which diseases and invertebrate pests may be detected within and around crops. This may be using one, or a combination of, visual, chemical, audio, or other data, which can often provide information about invertebrate and pathogenic species at an earlier stage than is possible through field crop observations. Most currently available technology consists of individual or combined sensors linked remotely to a cloud database, which can be placed in or near to the crop (Rustia et al. 2022). These collect data on target species is transferred to the user through a dedicated user interface. Depending on the associated support provided, this may include some additional guidance or alert to the grower where potential risks to the crop are detected. Depending on the number of these devices, their layout, and the frequency of data collection, these can provide increasing levels of precision in mapping risks to a crop, and so enable more precise management practices to be implemented. Where a disease or invertebrate pest, or reliable markers associated with them, can be detected at a small scale in a reliable way, this can be connected to automated management practice to apply targeted control measures.

A further advantage of utilizing automated detection systems is that as the range of species which can be detected grows, users will be able to track multiple risks to crops concurrently. This also enables improved tracking of a potentially new risk to a region, which the grower may not be otherwise alerted to, and to improved tracking of beneficial invertebrates.

4.4 Precision and Holistic Management of Pests and Diseases in Horticultural Crops

As growers work with their advisors and wider stakeholders to reduce the quantities and associated hazards of pesticides, a more advanced concept of IPM is needed that accounts for the wider implications of management practices. This 'Holistic' approach to IPM is built on 5 pillars of crop protection (Riemens et al. 2022; IPMWORKS consortium 2023):

Pillar 1: Agricultural landscapes with diverse semi-natural habitats
Pillar 2: Cropping systems designed to manage pests, weeds and diseases
Pillar 3: Optimized decision making guiding operational and strategic IPM choices
Pillar 4: Preferential use of non-chemical control options
Pillar 5: Increased efficiency of treatments

Wherever possible, all these pillars should be included in the design of crop management systems. In horticultural crops under precision management, the application of holistic IPM should be given priority over and implemented in advance of any precise management techniques. Preference of non-chemical control options, such as selection of crop cultivars more resistant to, or tolerant to pest damage, may require a different agronomic approach to that of more vulnerable varieties. In some cases, this may reinforce or remove the benefits of using precision techniques. The five pillars of holistic IPM provide a robust framework for advancing IPM strategies, but growers and advisors are left with practical challenges of implementation. As

Finch and Collier (2000) explain, without a large workforce it is difficult to monitor all vegetable fields for all potential pest infestations within the required timeframe to make effective management decisions. Without the ability to effectively monitor crops, decisions are weighted in favour of risk aversion rather than risk management, perpetuating reliance on pesticides. Investment in, and associated commitment to the use of, precision technology without its incorporation into a holistic IPM strategy can increase reliance on pesticides, albeit at a lower level, rather than supporting transition to true pesticide independence.

Growers and advisors require greater support in selection and implementation of preventative, long term approaches to pest management. This includes demonstrations of alternative approaches in practice, improved access to and support of effective DSS, and guidance in how to select, test, adapt, adopt, and integrate alternative approaches on farm.

4.5 Improving Adoption of Precision Technologies

Pest management presents some of the greatest challenges in agronomy, as the repercussions of poor control can lead to significant financial losses (Ramsden et al. 2017). Applying pesticides when the actual risk of pest damage is low can have chronic impacts on crop protection, such as driving selection of pesticide resistant species, and damaging naturally occurring populations of non-target organisms. Conversely, farmers can experience substantial economic losses where infestations are left unchecked and go on to cause widespread crop damage. Forecasts and detection systems as part of a precision horticulture approach can help avoid these problems but associated tools need to be sufficiently reliable and transparent for uptake to be widespread (Brewer and Goodell 2012; Rossi et al. 2014a).

Uptake of DSS is relatively high in horticultural cropping systems, compared with other agricultural sectors. Horticultural crops have generally higher value, and that value is impacted by pest activities to a greater extent than in other sectors. Horticultural pests can also be more challenging to manage, particularly in perennial crops where crop rotation is not possible. Development of tools for precision management of horticultural pests is well established, and usually develops from long standing collaborations between researchers and growers, in part due to the low tolerance to pesticide residues in the final product. While uptake is high compared with other sectors, the actual proportion of growers implementing precision IPM techniques remains low (Marinko et al. 2023). Uptake has been limited by various factors, associated with either the technical development and/or availability of systems, or the way in which farmers and their advisors are involved during and after development (Eastwood et al. 2019; Mackrell et al. 2009). Well researched systems are often rejected by potential users due to a lack of structured engagement, leading to a lack of trust or understanding of the outputs. Ultimately, this is driven by a failure to effectively engage with users and incorporate user characteristics into support systems, despite evidence of their influence in uptake and interpretation

(Aubert et al. 2012). Three considerations for improving application of precision technology as part of holistic IPM strategies include, i) balancing the technical function of a given tool against its practical application, ii) ensuring that the benefits and restrictions of a given tool have been openly validated and demonstrated, and iii) ensuring that the users of precision horticulture technologies have the information and resources to assess the impact of those tools for themselves.

Balancing Function with Useability

Most of the technological constraints identified in the 1990s as potential obstacles to the adoption of precision horticulture have since been alleviated by wider access to personal computers, the internet, and web-based services (Jones et al. 2010; Ascough et al. 2010), however some constraints still exist. Notably, developers have too often focused too much of their attention on implementing advanced technological features, while lower priority was given to ensuring Precision Technologies' credibility and their capacity to be integrated within the decision-making process (Rossi et al. 2019). Farmers and growers are already burdened by the diversity and depth of knowledge required to consistently produce economically viable crops. IPM measures are often complex and require an advanced understanding of inter-connected species ecology in order to be implemented effectively (Deguine et al. 2021). Suboptimal implementation of IPM measures can lead to disappointing results, which discourages farmers from continuing to follow this approach. This is a distinct disadvantage of IPM in comparison to conventional pesticide use, which requires less attention from the farmer and provides more consistent results (Alyokhin et al. 2015). Farmers and growers that make decisions in partnership with their agronomist, and with support from tool developers, will be able to best mitigate this risk.

Validation and Demonstration

A key aspect to be considered when selecting a precision horticulture technology, in particular a warning system and/or a DSS, is the extent to which the tool has been calibrated and validated in the proposed region of application. The calibration of a tool or model is the practice of estimating its parameters by solving the model's equations using observational data for both dependent and independent variables. The aim is to minimize the deviation between predicted and observed data. Once the calibration procedure proves satisfactory, the model must be evaluated in its capacity to provide reliable predictions of the biological processes it addresses. This procedure is called model validation and consists of comparing model outputs against field data (data must be different from those used in model calibration). Validation aims to demonstrate that the model can replicate real observations under different biotic and/or abiotic conditions (with respect to those used for model calibration) and guarantees the reliability of the model, as well as the possibility to generalize the model's outputs under different contexts (also called robustness of the model). In order to provide valuable information to users and facilitate the adoption of a precision technology, validation should also include a comparison of different models simulating the same process, when available (Caffi et al. 2007). The validation process is often performed by the model's developers because it requires thorough knowledge about the model including the modelling approach implemented, the equations used, and the biological processes represented.

Calibration and validation can be time-consuming and difficult to perform, especially when the precision technology accounts for multiple variables. Techniques to estimate parameter values (based on stochastic approaches and Bayesian methods) can facilitate precision technology calibration and validation (Gilioli et al. 2009). Lack of extensive validation affects a model's usefulness, limits its potential use, and it is questionable from an economic and ethical standpoint. Errors can result in the inaccurate estimation of pest's presence and abundance, and the misdiagnosis of pests. Unfortunately, models may be published and released to the market without a proper validation, essentially because there are no shared "minimum requirements" to be satisfied in order to consider a model as correctly validated (Rossi et al. 2019). In order to balance both the need to protect the benefits of robust, calibrated and validated systems, and support access to informative novel systems that may be adapted, growers should review the source and limitations of each system and consider these if and when consulting and interpretating them. Crucially, growers should endeavour to make their own assessment of the benefits of a given tool on their farm, before committing to widespread application.

Grower Led Assessment of Results

Growers can be reluctant to implement many aspects of precision pest management as part of integrated pest management strategies as the set-up costs and time investment needed can be high, while the benefits may not immediately translate into improved economic performance. Growers may also receive conflicting messages and information from different sources, making it difficult to identify, understand and apply the information they need. For these reasons, grower-led self-assessment of novel approaches is an important part of transition towards more advanced strategies, usually undertaken on part of one field, often in collaboration with researchers, advisors, and other farmers, comparing their standard practice to the alternative approach. A good way to compare and demonstrate different IPM strategies is to carry out two or more within the same field, in the same crop, under the same conditions. GPS and other modern technologies, along with thorough trial protocols, can make farm trialling straightforward and routine. Decisions and innovations can then become thoroughly validated and tailored to real farming conditions.

4.6 Discussion and Conclusion

Market demand for low costs, low pesticide residues, and no invertebrate or disease symptoms in the final product poses a real challenge and can limit which cultivars can be grown by farmers, in turn limiting the selection of pest resistant cultivars (Young et al. 2022). Spatial and temporal variability in populations of invertebrates and plant disease epidemics is often closely related to variability in weather conditions. As climates change and crop/variety selections shift, severe epidemics can occur in areas where a given invertebrate or disease may not have been traditionally considered a risk to productivity. The perceived risk of adoption of a fully integrated pest management approach, including application of precision technologies, in comparison to traditional pesticide use, is possibly the largest barrier to uptake.

Farmers and growers need assurance that the shift in management will not compromise crop yield, quality, or market value, or require significant investment and change to existing processes that would impact the economic viability of production. Many of the perceived risks of changing strategy relate to the high short-term costs of building an IPM strategy, with long term benefits, compared to low costs and short-term benefits of more conventional pesticide application programmes (Brewer and Goodell 2012; Doonan 2017). Economic concerns are widely acknowledged to be a critical barrier to adoption of IPM. However, potential economic benefits associated with longer term resilience and robustness to production, are also often given as a motivation for uptake (Glare et al. 2019; Khandelwal et al. 2016; Andert et al. 2018;).

Despite the challenges of complex implementation, and market driven rather than yield based profits (Beres et al. 2011; Buurma and van der Velden 2017), there is an appetite among horticultural producers to reduce reliance on pesticides through precision technologies. Vegetable growers who were practicing IPM reported a 35% reduction in their overall pesticide use by using more targeted pesticide applications (Horrocks et al. 2018), largely through improving growers understanding of the role of natural predators (Tang et al. 2010) and the use of thresholds and decision tools to guide pesticide use (Walker et al. 2009). The combination of site-specific weather data, monitoring reports, and advice from a DSS, enables growers and their advisors to improve crop protection, for instance in vineyards against downy mildew by modulating the frequency and timing of fungicide applications based on disease risk (Rossi et al. 2023). However, there remains a gap between applied research and practical application, due to a lack of published research and/or the accessibility of research to practitioners (Gent et al. 2013), especially where researchers are rewarded for creation of papers within an academic setting, rather than providing practical solutions to real, on-farm issues. This is a supply chain issue as there needs to be research to link up the scientific papers and actions taken on farm (applied research). More farmer-to-farmer initiatives are also needed to facilitate and share on-farm innovation and provide support groups to encourage successful IPM adoption (Bürger et al. 2012).

As further precision technologies are developed, and these are better validated and demonstrated to growers, uptake may increase. This could enable improved targeting of preventative and responsive management of infestations, and reduce the overall quantities of pesticides required. This advancement requires engagement from all stakeholders, including researchers, developers, agronomists, farmers and growers, to ensure that the systems are both effective, affordable and implementable, and are designed to fit individual farm strategies without compromising other aspects of production.

4.7 Competing Interests

The authors have no conflicts of interest to declare that are relevant to the content of this chapter. This chapter was produced as part of two projects received funding from the European Research Council (ERC) under the European Union's Horizon 2020 research and innovation programme; IPM Decisions (grant agreement 817617) and IPMWORKS (grant agreement 101000339). Furiosi M. conducted this study within the Doctoral School of the Agro-Food System (Agrisystem) of Università Cattolica del Sacro Cuore (Italy).

References

Alyokhin A, Mota-Sanchez D, Baker M, Snyder WE, Menasha S, Whalon M, Dively G, Moarsi WF (2015) The Red Queen in a potato field: integrated pest management versus chemical dependency in Colorado potato beetle control. Pest Manag Sci 71(3):343–356. https://doi.org/10.1002/ps.3826

Andert S, Buerger J, Mutz JE, Gerowitt B (2018) Patterns of pre-crop glyphosate use and in-crop selective herbicide intensities in Northern Germany. Eur J Agron 97:20–27

Ascough II, James C, McMaster GS, Dunn GH, Andales AA (2010) The GPFARM DSS for agroecosystem sustainability: past, future, and lessons learned. In: International congress on Environmental Modelling and Software, p 313

Aubert BA, Schroeder A, Grimaudo J (2012) IT as enabler of sustainable farming: an empirical analysis of farmers' adoption decision of precision agriculture technology. Decis Support Syst 54(1):510–520. https://doi.org/10.1016/j.dss.2012.07.002

Bajwa WI, Kogan M (2002) Compendium of IPM definitions (CID) what is IPM and how is it defined in the worldwide literature? IPPC Publ 998:1–14

Barzman M, Bàrberi P, Birch ANE, Boonekamp P, Dachbrodt-Saaydeh S, Graf B et al (2015) Eight principles of integrated pest management. Agron Sustain Dev 35:1199–1215

Beres BL, Cárcamo HA, Yang RC, Spaner DM (2011) Integrating spring wheat sowing density with variety selection to manage wheat stem sawfly. Agron J 103(6):1755–1764

Blackmore S, Godwin R, Fountas S (2003) The analysis of spatial and temporal trends in yield map data over six years. Biosyst Eng 84:455–466. https://doi.org/10.1016/S1537-5110(03)00038-2

Bowden J, Cochrane J, Emmett BJ, Minall TE, Sherlock PL (1982) A survey of cutworm attacks in England and Wales, and a descriptive population model for Agrotis segetum (Lepidoptera: Noctuidae). Ann Appl Biol 102:29–47. https://doi.org/10.1111/j.1744-7348.1983.tb02663.x

Brewer MJ, Goodell PB (2012) Approaches and incentives to implement integrated pest management that addresses regional and environmental issues. Annu Rev Entomol 57:41–59. https://doi.org/10.1146/annurev-ento-120709-144748

Bürger J, de Mol F, Gerowitt B (2012) Influence of cropping system factors on pesticide use intensity–a multivariate analysis of on-farm data in North East Germany. Eur J Agron 40:54–63

Buurma JS, van der Velden NJA (2017) New approach to integrated Pest management research with and for horticulture. A vision from and beyond economics. Crop Prot 97:94–100. https://doi.org/10.1016/j.cropro.2016.11.013

Caffi T, Rossi V, Cossu A, Fronteddu F (2007) Empirical vs. mechanistic models for primary infections of Plasmopara viticola. EPPO Bull 37:261–271

Cardim Ferreira Lima M, Damascena de Almeida Leandro ME, Valero C, Pereira Coronel LC, Gonçalves Bazzo CO (2020) Automatic detection and monitoring of insect pests—a review. Agriculture 10:161. https://doi.org/10.3390/agriculture10050161

Chaplin-Kramer R, O'Rourke ME, Blitzer EJ, Kremen C (2011) A meta-analysis of crop pest and natural enemy response to landscape complexity. Ecol Lett 14:922–932. https://doi.org/10.1111/j.1461-0248.2011.01642.x

Collier R, Mazzi D, Schjoll AF, Schorpp Q, Thoming G, Johansen TJ, Meadow R, Mayling NV, Cortesero A, Vogler U, Gaffney MT, Hommes M (2020) The potential for decision support tools to improve the management of root-feeding fly pests of vegetables in Western Europe. Insects 11(6):369. https://doi.org/10.3390/insects11060369

Collier R, Cortesero A, Meyling N, Hommes M, Vogler U, Schorpp Q, Szikora T, Gaffney M, Johansen T, Meadow R, Trdan S (2021) Integrated control of root-feeding fly larvae infesting vegetable crops (FlyIPM). Danish Environmental Protection Agency, Denmark. Pesticide Research No. 195 ISBN: 978-87-7038-220-5

Deguine JP, Aubertot JN, Flor RJ et al (2021) Integrated pest management: good intentions, hard realities. A review. Agron Sustain Dev 41:38. https://doi.org/10.1007/s13593-021-00689-w

Doonan H (2017) Uptake of integrated pest management in the UK arable sector. Outlooks Pest Manage 28(1):29–32

Eastwood C, Ayre M, Nettle R, Dela Rue B (2019) Making sense in the cloud: farm advisory services in a smart farming future. NJAS Wageningen J Life Sci 90–91:100298. https://doi.org/10.1016/j.njas.2019.04.004

Esbjerg P, Sigsgaard L (2019) Temperature dependent growth and mortality of *Agrotis segetum*. Insects 10(1):7. https://doi.org/10.3390/insects10010007

FAO (2024) Sustainable development goals. Available at https://www.fao.org/sustainable-development-goals-helpdesk/en. Accessed 14 Oct 2024

Finch S, Collier RH (2000) Integrated pest management in field vegetable crops in northern Europe — with focus on two key pests. Crop Prot 19(8–10):817–824. https://doi.org/10.1016/S0261-2194(00)00109-5

Fraser I, Sharma A, Bailey A (2011) Multiple adoption of Pest management technologies in UK cereal farming. In: 55th annual AARES national conference, Melbourne, Victoria. Australian Agricultural and Resource Economics Society

Gebbers R, Adamchuk VI (2010) Precision agriculture and food security. Science 327:828–831. https://doi.org/10.1126/science.1183899

Gent DH, Mahaffee WF, McRoberts N, Pfender WF (2013) The use and role of predictive systems in disease management. Annu Rev Phytopathol 51(1):267–289

Gilioli G, Pasquali S, Morlacchi P, Lozzia GC, Baumgärtner J (2009) Development of models for the analysis of the dynamics and management of plant-phytophagous-predator systems. In: Proceedings of the V study days on models for plant protection

Glare T, McKinnon A, Gwynn R, Brownbridge M (2019) Advances in microbial control in integrated pest management: entomopathogenic fungi. In: Integrated management of insect pests. Burleigh Dodds Science Publishing, pp 549–592

Hasanaliyeva G, Ammour MS, Yaseen T, Rossi V, Caffi T (2022) Innovations in disease detection and forecasting: a digital roadmap for sustainable management of fruit and foliar disease. Agronomy 12(7):1707. https://doi.org/10.3390/agronomy12071707

Horrocks A, Horne PA, Davidson MM (2018) Demonstrating an integrated pest management strategy in forage-and seed-brassica crops using a collaborative approach. N Z Plant Prot 71: 112–120

IPM Decisions (2022) IPM decisions platform. Available at: https://www.platform.ipmdecisions.net/. Accessed 14 Oct 2024

IPMWORKS consortium (2023) Policy recommendation for the Sustainable Use of Pesticides Regulation (SUR), 12 pages. https://doi.org/10.57745/XWADQJ

Isard SA, Russo JM, Magarey RD, Golod J, VanKirk JR (2015) Integrated pest information platform for extension and education (iPiPE): progress through sharing. J Integr Pest Manage 6(1):15. https://doi.org/10.1093/jipm/pmv013

Jones O, Kirwan J, Morris C, Buller H, Dunn R, Hopkins A, Whittington F, Wood J (2010) On the alternativeness of alternative food networks: sustainability and the co-production of social and ecological wealth. In: Fuller D, Jonas AE, Lee R (eds) Interrogating alterity: alternative economic and political spaces. Ashgate Publishing, Ltd, Farnham, pp 95–109

Khandelwal N, Barbole RS, Banerjee SS, Chate GP, Biradar AV, Khandare JJ, Giri AP (2016) Budding trends in integrated pest management using advanced micro-and nano-materials: challenges and perspectives. J Environ Manag 184:157–169

Kumar D, Kumari P, Kamboj R et al (2022) Entomopathogenic nematodes as potential and effective biocontrol agents against cutworms, Agrotis spp.: present and future scenario. Egypt J Biol Pest Control 32:42. https://doi.org/10.1186/s41938-022-00543-5

Kunjwal N, Srivastava RM (2018) Insect pests of vegetables. In: Pests and their management. Springer, Singapore, pp 163–221

Leskinen M, Markkula I, Koistinen I, Pylkkö P, Ooperi S, Siljamo P, Ojanen H, Raiskio S, Tiilikkala K (2011) Pest insect immigration warning by an atmospheric dispersion model, weather radars and traps. J Appl Entomol 135(1–2):55–67. https://doi.org/10.1111/j.1439-0418.2009.01480.x

Mackrell D, Kerr D, von Hellens L (2009) A qualitative case study of the adoption and use of an agricultural decision support system in the Australian cotton industry: the socio-technical view. Decis Support Syst 47(2):143–153. https://doi.org/10.1016/j.dss.2009.02.004

Magarey RD, Isard SA (2017) A troubleshooting guide for mechanistic plant Pest forecast models. J Integr Pest Manage 8:1. https://doi.org/10.1093/jipm/pmw015

Magarey RD, Travis JW, Russo JM, Seem RC, Magarey PA (2002) Decision support systems: quenching the thirst. Plant Dis 86:4–14. https://doi.org/10.1094/PDIS.2002.86.1.4

Mahlein A (2016) Plant disease detection by imaging sensors—parallels and specific demands for precision agriculture and plant phenotyping. Plant Dis 100(2):241–251. https://doi.org/10.1094/PDIS-03-15-0340-FE

Marinko J, Ivanovska A, Marzidovšek M, Ramsden M, Debeljak M (2023) Incentives and barriers to adoption of decision support systems in integrated pest management among farmers and farm advisors in Europe. Int J Pest Manag:1–18. https://doi.org/10.1080/09670874.2023.2244912

Nitzko S, Bahrs E, Spllier A (2022) Pesticide residues in food and drinking water from the consumer's perspective: the relevance of maximum residue levels and product-specific differences. Sustainable Prod. Consumption 30:787–798. https://doi.org/10.1016/j.spc.2022.01.016

Pierpaoli E, Carli G, Pignatti E, Canavari M (2013) Drivers of precision agriculture technologies adoption: a literature review. Procedia Technol 8:61–69. https://doi.org/10.1016/j.protcy.2013.11.010

Pimentel D (2005) Environmental and economic costs of the application of pesticides primarily in the United States. Environ Dev Sustain 7:229–252. https://doi.org/10.1007/s10668-005-7314-2

Preti M, Verheggen F, Angeli S (2021) Insect pest monitoring with camera-equipped traps: strengths and limitations. J Pest Sci 94:203–217. https://doi.org/10.1007/s10340-020-01309-4

Ramsden MW, Kendall SL, Ellis SA, Berry PM (2017) A review of economic thresholds for invertebrate pests in UK arable crops. Crop Prot 96:30–43

Riemens M, Sønderskov M, Moonen A, Storkey J, Kudsk P (2022) An integrated weed management framework: a pan-European perspective. Eur J Agron 133:126443. https://doi.org/10.1016/j.eja.2021.126443

Rossi V, Giosuè S, Caffi T (2010) Modelling plant diseases for decision making in crop protection. In: Oerke EC, Gerhards R, Menz G, Sikora RA (eds) Precision crop protection – the challenge and use of heterogeneity. Springer, pp 241–258

Rossi V, Caffi T, Salinari F (2012) Helping farmers face the increasing complexity of decision-making for crop protection. Phytopathol Mediterr 51:457–479

Rossi V, Salinari F, Poni S, Caffi T, Bettati T (2014a) Addressing the implementation problem in agricultural decision support systems: the example of vite.net?? Comput Electron Agric 100:88–99. Available at: http://linkinghub.elsevier.com/retrieve/pii/S0168169913002536. Accessed 29 Nov 2013

Rossi V, Caffi T, Legler SE, Carotenuto E, Bigot G, Brunelli A et al (2014b) Large-scale application of a web-based decision support system for sustainable viticulture. IOBC/WPRS Bull 105:129–136

Rossi V, Sperandio G, Caffi T, Simonetto A, Giolioli G (2019) Critical success factors for the adoption of decision tools in IPM. Agronomy 9:1–19

Rossi V, Caffi T, Salotti I et al (2023) Sharing decision-making tools for pest management may foster implementation of integrated Pest management. Food Sec 15:1459–1474. https://doi.org/10.1007/s12571-023-01402-3

Rustia DJA, Lin-Ya C, Chen-Yi L, Ya-Fang W, Sheng-Kua C, Jui-Yung C, Ju-Chun H, Ta-Te L (2022) Towards intelligent and integrated pest management through an AIoT-based monitoring system. Pest Manag Sci 78(10):4288–4302. https://doi.org/10.1002/ps.7048

Shepherd M, Turner JA, Small B, Wheeler D (2018) Priorities for science to overcome hurdles thwarting the full promise of the 'digital agriculture' revolution. J Sci Food Agric 100(14): 5083–5092

Tang S, Tang G, Cheke RA (2010) Optimum timing for integrated pest management: modelling rates of pesticide application and natural enemy releases. J Theor Biol 264(2):623–638

Tilman D, Balzar C, Hill J et al (2011) Global food demand and the sustainable intensification of agriculture. Proc Natl Acad Sci U S A 108:20260–20264

Walker JTS, Park NM, Clothier BE, Manktelow DWL, Van Den Dijssel CW, Hodson AJ, Barley MJ, Hodson-Kersey LR (2009) Progress in pesticide risk reduction in New Zealand horticulture. N Z Plant Prot 62:321–327

Young C, Cook S, Wedgwood E, Bennison J, Bartel E, Blake J, Huckle A, Allen J, Godfrey K, Eyre C, Butters L, Rees HJ, Bowsher-Gibbs M, Creissen HE (2022) Review and guidance for integrated management of economically significant weeds, pests and diseases in a range of horticultural and other edible field crops. Agricultural And Horticultural Development Board (AHDB). https://projectbluearchive.blob.core.windows.net/media/Default/Research%20Papers/Horticulture/CP%20211_Final%20report_IPM%20non-broadacre%20crops.pdf

Chapter 5
Weed Management—Identification and Treatment

Panagiotis Kanatas, Ioannis Gazoulis, Alexandros Tataridas, Anastasia Tsekoura, and Ilias Travlos

Abstract Precision horticulture and site–specific weed management (SSWM) practices based on remote sensing techniques have a great potential to contribute to the development of intelligent, effective, and more environmentally friendly weed management systems. The most crucial and challenging task to accomplish is correct weed identification in the field. Non–imaging and imaging ground-based sensors can be used for this purpose. Non–imaging weed identification systems include spectrometric sensors, optoelectronic sensors, UV–induced fluorescence sensors, LiDAR sensors, and ultrasonic sensors. As for imaging systems, RGB–imaging is the most widespread remote sensing technique to detect weed patches in the field while the combination of imaging and non–imaging sensors to obtain hyperspectral data is also very promising. Unmanned aerial vehicles (UAV) equipped with smart sensors can also be used to collect information on weed flora composition in horticultural systems. Once data have been captured, they are processed with machine learning algorithms to distinguish between either weeds and bare soil or between weeds and crops. Data can be used to create weed maps to guide precise mechanical weed control operations and site-specific herbicide applications. Sensor–based mechanical weed control can be achieved with semi-autonomous tractor–drawn implements and autonomous robots. To perform site-specific herbicide applications, a reliable weed identification system, a processing unit, and precision weed control implements are needed. Precision horticultural methodologies can result in significant herbicide savings while sensor-based methods can be used to evaluate herbicide efficacy. However, the use of site-specific weed management tools by growers remains

---

P. Kanatas (✉)
University of Patras, GR30200, Mesolonghi, Greece

I. Gazoulis · A. Tataridas · I. Travlos (✉)
Agricultural University of Athens, GR11855, Athens, Greece
e-mail: travlos@aua.gr

A. Tsekoura
Benaki Phytopathological Institute, GR14561, Kifisia, Greece

P. Newell Price et al. (eds.), *Precision Agriculture in Field Horticulture*, Progress in Precision Agriculture, https://doi.org/10.1007/978-3-031-96534-0_5

sporadic. To increase adoption rates, technical experts need to establish stronger links with growers and their advisers to familiarize them with the concepts of precision horticulture and site-specific weed management.

Keywords Precision horticulture · Weed detection · Non-imaging sensors · Imaging sensors · Machine learning algorithms · Site-specific · Mechanical weed control · Herbicide application

5.1 Introduction

Weeds are competitive and adaptable to their environment (Chauhan 2020) and are considered one of the most significant constraints to agricultural production (Gharde et al. 2018). Oerke and Dehne (2004) estimated yield reductions due to weed competition of 5%, 10% and 25% for the most, less and least developed countries, whilst a later study by the same author (Oerke 2006) cited a higher figure of 34% yield loss in global agricultural production. In addition, weeds can have a negative impact on crop quality, act as hosts for pests and diseases, and generally increase the cost of cultivation process (Zimdahl 2018).

In most countries around the world, herbicide application is the dominant option to control weeds and achieve higher crop yields. Moreover, herbicides are cost–effective and their use is associated with lower levels of soil erosion, fuel use, and greenhouse gas emissions (Gianessi 2013). However, overreliance on herbicides with similar modes of action has resulted in the evolution of herbicide–resistant weeds and in the selection of several herbicide resistant populations. Globally, more than 500 unique cases of herbicide resistance have been reported (Heap 2020). Other issues associated with imprudent herbicide use may include crop injury, injury to non–target vegetation, the existence of residues in the soil environment and water table, toxicity to non–target organisms and concerns for human health and safety (Li et al. 2003; Cox 2006; Pot et al. 2011). As a result, there is an urgent need to shift from overreliance on herbicides and conventional practices to alternative, innovative, and sustainable weed management systems.

Precision horticulture and site–specific weed management (SSWM) practices based on remote sensing techniques have great potential to contribute to the development of intelligent, effective, and more environmentally friendly weed management systems (Roslim et al. 2021). The most crucial and difficult task to accomplish is the correct weed detection in the field (Thorp and Tian 2004; López-Granados 2011; Su 2020a, b). Non–imaging and imaging ground based sensors can be used for weed identification purposes. Non–imaging weed identification systems include spectral reflectance sensors i.e., spectrometric and optoelectronic sensors, UV–induced fluorescence sensors, and distance sensors i.e., LiDAR and ultrasonic sensors (Peteinatos et al. 2014; Fernández-Quintanilla et al. 2018; Roslim et al. 2021). As for imaging systems, RGB–imaging is the most widespread remote sensing technique to detect weed patches in the field while the combination of imaging and non–imaging sensors to obtain hyperspectral data is also very promising (Su 2020a, b; Roslim et al. 2021). In addition, unmanned aerial vehicles (UAV) equipped with smart sensors can collect information on weed flora composition in horticultural areas (Malamiri et al. 2021; Mohidem et al. 2021). Once captured, data can be processed using machine learning algorithms that distinguish between weeds, crops, and bare ground (Su 2020a). Successful mapping of weeds on horticultural land can, then, precisely guide subsequent mechanical weed control treatments or herbicide applications (Christensen et al. 2009; López-Granados 2011; Gonzalez-de-Soto et al. 2016; Machleb et al. 2020).

The aim of this chapter is to review how weed identification can be carried out using precision horticulture methodologies. The performances of smart sensors belonging to each category that have been used for this purpose are presented according to evidence derived from the recent existing literature. In addition, we present case studies where remote sensing techniques have been used to guide mechanical weed control and for precise application of herbicides. Future challenges that need to be addressed to increase the adoption of precision horticulture methodologies for weed management are also discussed.

5.2 The Potential of Sensor–Based, Precision Horticulture Methodologies for Weed Identification

The Potential of Spectral reflectance Sensors for Weed Identification

Spectral reflectance sensors include spectrometric and optoelectronic sensors (Fernández-Quintanilla et al. 2018). Spectrometric sensors measure reflection intensities at wavelengths within a range of the electromagnetic spectrum, ranging from ultraviolet (UV), visible light, to near–infrared (NIR) (Peteinatos et al. 2014). Given that the green leaves of actively growing weeds are characterized by a high reflectance in the green and NIR and a low reflectance in the red and blue spectra, spectrometers can differentiate weeds from bare soil (Marchant and Onyango 2003; Kavdr 2004).

Optoelectronic sensors focus on few specific spectral bands and particularly on bands in the red/near–infrared (R/NIR) spectrum (Scotford and Miller 2005). As for practical examples of how these sensors operate, Dammer and Wartenberg (2007) provide specific details.

In general, spectral reflectance sensors can discriminate between plant presence and absence by measuring indices correlated with plant coverage values (Felton and McCloy 1992). The ability of such sensors to differentiate weeds from soil enables weed detection in areas between crop rows in wide–spaced row crops. For example, Sui et al. (2008) evaluated an active WeedSeeker® PhD600 optical sensor for weed mapping in cotton (*Gossypium hirsutum* L.) spaced 97–cm apart and found a good relationship between sensor data and actual weed presence between. Dammer and

Wartenberg (2007) performed a four–year field evaluation of an optoelectronic sensor in fields cultivated with cereals and legumes. The evaluations were carried out on the tramlines of these narrow–spaced crops and provided recommendations for site–specific weed control. In addition to weed identification between crop rows, spotting weed patches before crop emergence could be useful for weed management if a false or stale seedbed is prepared (Kanatas et al. 2020a, b). However, a strong limitation of this category of sensors is their inability to discriminate between weeds and crops and also to identify weed species composition in a particular patch (Peteinatos et al. 2014; Fernández-Quintanilla et al. 2018).

To deal with this major challenge, the use of spectral reflectance measurements should be integrated with the use of machine learning models. Examples of such models are summarized in a recent review article by Su (2020a, b): cluster analysis (CA), discriminant analysis (DA), convolutional neural networks (CNN), artificial neural networks (ANN), support vector machines (SVM), and soft independent modelling of class analogy (SIMCA), random forest (RF) and others.

De Souza et al. (2020) used a Fieldspec 4 Std–Res spectroradiometer able to make readings in the spectral range from 450 to 2500 nm. The tested models SIMCA, and RF simplified the Vis–NIR spectrum into only four bands of interest (500–550 nm; 650–750 nm; 1300–1450 nm; and 1800–1900 nm) resulting in 97% successful discrimination between sugarcane (*Saccharum officinarum* L.) and a mixed stand of grass and broadleaf weeds that were infesting the crop. In maize (*Zea mays* L.), a spectral–reflectance–based Plant Discrimination Unit (PDU) was used to collect laser beams reflected off the investigated plants and background (Akbarzadeh et al. 2018). The use of the SVM method in combination with spectral reflectance measurements increased discrimination accuracy between maize and broadleaf weeds by 27% compared to conventional practices. In winter oilseed rape (*Brassica napus* L.), an Analytical Spectral Device was used to capture spectral data of soil, crop plants and weeds within the 350–2500 nm wavelength range (Wei et al. 2014). After capturing spectral data, these researchers used the elimination of uninformative variables (UVE) and successive projections algorithm (SPA) models and selected five effective wavelengths (672, 757, 897, 1117, and 1722 nm). The final result was to achieve more than 98% weed–crop discrimination through employing canonical discriminant analysis (CDA) on the five selected wavelengths. In a white cabbage (*Brassica oleracea* L. var. *capitata*) field infested with barnyardgrass (*Echinochloa crus–galli* L.), green foxtail (*Setaria viridis* L.), goosegrass (*Eleusine indica* L.), and goosefoot (*Chenopodium quinoa* Willd.), spectrometer measurements were subjected to principal component analysis (PCA) to select feature wavelengths (Deng et al. 2016), and, Bayesian discriminant analysis successfully classified 2 cabbage cultivars and all 5 weed species with an accuracy of 84.3%. Similarly, Kokalj et al. (2017) combined spectrometer measurements with cluster analysis (CA) and achieved complete classification between rocket salad [*Eruca vesicaria* (L.) Cav.] and groundsel (*Senecio vulgaris* L.). Although the results of these studies are very promising, many have been conducted under controlled conditions and, further research should be conducted to optimize weed identification using spectral reflectance sensors under the field conditions where variability of weather conditions is an obstacle for capturing reliable spectral data (Peteinatos et al. 2014).

The Potential of UV–Induced Fluorescence Sensors for Weed Identification

UV–induced fluorescence sensors measure the emitted fluorescent radiation of leaves after exposing plants with a radiation of visible or UV light for a specific amount of time and intensity (Peteinatos et al. 2014). It should be noted that the fluorescent wavelength is longer than the excitation light wavelength, and emitted in a small amount of time after receiving light (Fernández-Quintanilla et al. 2018). The ultraviolet (UV) induced fluorescence of the plants appears as a promising technique for real–time weed detection (Liu and Bruch 2020). UV radiation stimulates blue–green fluorescence (450 nm) of the leaves. Chlorophyll *a* and *b* emit fluorescence in the red to far–red range of the spectrum (680–700 and 735–750 nm) (Krause and Weis 1991). The ratio of blue–green fluorescence to chlorophyll *a* fluorescence has been shown to be highly correlated to plant species and can be used for plant identification (Cerovic et al. 1999).

Panneton et al. (2010) showed that measuring UV–induced fluorescence in the blue–green spectrum under controlled greenhouse conditions could be used to effectively discriminate between maize and broadleaf weeds. In this study, fluorescence was collected by a fiber–optic bundle (Oriel 77532) and transported to an Oriel MS125 spectrograph. A partial least squares discriminant analysis (PLS–DA) was performed and two large bands were selected in the blue–green fluorescence zone (400–425 nm and 425–490 nm). A linear discriminant analysis followed and showed sufficient accuracy. However, this process could not be used to sufficiently accurately differentiate between maize and grass weeds. In addition, Longchamps et al. (2010) used UV–induced fluorescence for weed–crop discrimination in maize. An Oriel MS 125 spectrograph was again used. A linear discriminant analysis was applied on fluorescence data derived from maize, grass, and broadleaf weeds. A recognition rate of 91.8% was achieved; however, these results were obtained under controlled greenhouse conditions and not in the field. Tyystjärvi et al. (2011) reported results from another pot experiment conducted on maize and on barley (*Hordeum vulgare* L.). Pots were placed outdoors and subjected to natural weather conditions, and six weed species were included. A PAM–101 fluorometer was used for data collection. Under this concept known as fluorescence fingerprinting, leaves are treated with a series of different colors and intensities of light, and changes in chlorophyll *a* fluorescence yield are recorded to produce a fluorescence curve. In the study, a neural network algorithm (NNA) was used to process the produced fluorescence curves and correctly classified approximately 86–97% of the field as crop or weed. Gerhards et al. (2012) used a MiniVegN sensor under both greenhouse and field conditions for weed mapping in wheat. Red and infrared far–red chlorophyll *a* fluorescence was induced by a red laser and when the laser beam hit a plant, fluorescence was induced and recorded. In cases where sensor moved with a steady speed, noxious weed species such as black–grass (*Alopecurus myosuroides* Huds.) were correctly discriminated from wheat with an accuracy that surpassed 70%. Although these sensors offer potential for weed mapping, further work needs to be carried out to validate the methodology under field conditions and in a wider variety of field and horticultural crops.

The Potential of Distance Sensors for Weed Identification

Distance sensors, including Light Detection and Ranging (LiDAR) sensors and ultrasonic sensors, can be used to provide reliable estimates of plant height and biomass (Legg and Bradley 2020; Bronson et al. 2021; Colaço et al. 2021; El-Naggar et al. 2021). Since they can provide accurate assessments of these particular plant growth parameters, they can contribute to weed mapping in horticultural crops. Although limitations exist as in every category of sensors, general advantages of distance sensors are that they are more cost effective than others and friendly for the user (Fernández-Quintanilla et al. 2018).

LiDAR sensors use a laser beam to measure the distance between the sensor and the targeted weeds emerging in the field (Peteinatos et al. 2014). There are two methods of measuring this distance. The first is based on measuring the time it takes for the laser pulse to travel between the emitter and the receiver of the sensor and be reflected by a particular weed individual. According to Ehlert et al. (2009), this is the 'time–of–flight' principle. The second method is based on measuring the phase difference between the emitted and reflected laser beam. These are the so–called 'phase–shift measurements' according to Rosell and Sanz (2012). The use of LiDAR sensors is classified as a laser imaging technique by Fernández-Quintanilla et al. (2018). In contrast, the use of ultrasonic sensors is classified as an ultrasound technique. The function of ultrasonic sensors for weed detection are exclusively based on the 'time–of–flight' principle and particularly on the measurement of reflected sound waves. An ultrasonic pulse receiver is used to measure the distance between the sensor and the targeted weeds (Machleb et al. 2020). In particular, the waves return to the receiver after impacting the surface of the targeted weeds, and the travel time of the acoustic signals are measured and transformed into voltage signals. The output voltages are then converted into distance values (Andújar et al. 2012).

Shahbazi et al. (2021) reported results from a study using LiDAR to identify weeds in a wheat (*Triticum aestivum* L.) field. The target was to estimate vegetation height and calculate distances between the sensor and the weeds infesting the crop. Distance calculations were based on the return times of the laser beams ('time–of–flight' principle). The correlations between LiDAR recordings and actual heights of both wild oat (*Avena fatua* L.) and annual sow thistle (*Sonchus oleraceus* L.) were noticeable ($R^2 = 0.99$). The results derived from the research by Andújar et al.

(2013a) were also promising. These authors used a Terrestrial Laser Scanner (TLS) sensor based on phase shift LiDAR technology to estimate vegetation height in a maize (*Zea mays* L.) field. The sensor was able to effectively detect johnsongrass [*Sorghum halepense* (L.) Pers.] patches infesting the maize crop since strong correlations ($R^2 = 0.86$) were observed between johnsongrass height and LiDAR measurements. Similar results were found for common cocklebur (*Xanthium strumarium* L.), but the detection of shorter weed species was not possible. These results suggest that LiDAR sensors are more suitable for identification of tall weed species which grow above the height of the crop canopy. In the next work by Andújar et al. (2013b), it was found that although weed species of lower plant height [barnyardgrass; *Echinochloa crus–galli* (L.) P.Beauv., red dead–nettle; *Lamium purpureum* L., and catchweed bedstraw; *Galium aparine* L.] were present in this maize field, LiDAR estimations were again significantly correlated with actual plant height values ($R^2 = 0.75$). However, the real progress was the ability to discriminate bare soil, maize plants, grass weeds and broadleaf weeds as well with an overall success of over 70%. Identifying grass weeds was challenging due to similarities with the crop but the authors denote that such problems do not exist for inter–row weed detections.

Regarding the contribution of ultrasonic sensors in the sector of weed identification, Andújar et al. (2011a) reported interesting results from studies in two maize fields infested with johnsongrass and broadleaf weeds. Johnsongrass plants were significantly taller than broadleaf weeds. Ultrasonic measurements showed strong relationships with johnsongrass height and biomass ($R^2 = 0.73$), while the performance of the sensor was lower on detecting broadleaf weeds. This study was one of the first highlighting ultrasonic sensors as a promising tool for weed identification in field crops grown in wide rows infested with weed species that form tall, erect canopies. Moreover, weed classification into pure stands of grasses, pure stands of broadleaves and mixed stands of grasses and broadleaves reached 73.4, 87.9, and 53.4% accuracy. The same equipment was used in another study to identify johnsongrass patches in maize at crop harvest, and to create maps predicting the spread of these patches in the field during the upcoming years (Andújar et al. 2011b). In another study, an ultrasonic sensor again provided accurate assessments of weed height, but the authors emphasized that the reliability of their method should be tested in earlier growth stages of maize (Rueda-Ayala et al. 2015).

In studies carried out in winter wheat, Andújar et al. (2012) found contrasting results on the effectiveness of ultrasonic sensors to detect weeds. A multiple regression model was run to estimate relationships between ultrasonic readings and actual weed/crop growth parameters. The first evaluation was carried out between wheat growth stages 11 and 13 (BBCH scale). Significant relationships were found between ultrasonic measurements and the density, height and biomass of grass weeds, the height and biomass of broadleaf weeds as well as regarding total weed coverage. However, the results of the second evaluations revealed no significant correlations between ultrasonic and actual observations for any of the studied parameters. The authors attributed these results to the growth rates of wheat and the narrower row spacing that is used for its establishment. These factors enable the crop to shade weeds rapidly making ultrasonic measurements difficult at later growth stages. Thus, they suggested that ultrasonic sensors should be used for weed detection in early crop growth stages in narrow–row spaced crops such as wheat while estimations can be quite accurate in later crop growth stages in wide–row spaced crops such as maize. Concerning other crops with different morphology than cereals Farooque et al. (2013) reported encouraging results in a wild blueberry (*Vaccinium angustifolium* Ait.) stand. The readings of the ultrasonic sensor were effectively used to create maps to monitor weed presence in the field and explain variations in the yield performance of the crop. Strong correlations ($R^2 = 0.98$) were also obtained between the height of the weed vegetation and the estimations of the sensor as shown by the linear regression analysis performed by other researchers in the same crop (Swain et al. 2009).

The Potential of RGB–Imaging Sensors for Weed Identification

The use of imaging sensors for weed detection is probably the most researched of the precision methods for weed detection (Brown and Noble 2005; Wang et al. 2019). RGB imaging refers to the use of an RGB digital camera equipped with three color filters (red, green & blue) to capture an image, mimicking the way the human eye perceives color (Su 2020a). The use of RGB sensors to acquire image data for plant species discrimination is well established and is one of the most frequently used ground–based sensors for weed–crop discrimination under real field conditions (Peteinatos et al. 2014; Wang et al. 2019; Su 2020b). Spectral features, biological morphology, visual textures, spatial contexts, and patterns present in digital images can be used to discriminate between weeds and crops (Slaughter et al. 2008; Liu and Bruch 2020). The RGB system of color cameras is known for its high spatial and limited spectral resolution; such imaging systems with high spatial resolution can achieve pixel resolutions of the order of a few millimeters or even smaller (Gai et al. 2020). Broadband reflectance images are acquired which are sufficient to effectively distinguish plants from bare soil based on spectral differences (Su 2020b). After segmenting the acquired image with broadband reflectance data, the features representing morphology or texture of plant canopy and plant leaf can be extracted for different plant differentiations (Su 2020a).

Burgos-Artizzu et al. (2009) used Nikon Coolpix 5700 and Sony DCR PC110E RGB cameras to obtain images of weeds infesting a barley (*Hordeum vulgare* L.) crop. Corn poppy (Papaver rhoeas L.) and sterile oat (*Avena sterilis* L. subsp. *ludoviciana*) were the dominant weed species. Each image covered an area of 0.51 m^2. A Case–Based Reasoning (CBR) approach was used for image processing. This concept resulted in correlation coefficients of almost 80% between image and actual data. A Nikon Coolpix 5400 RGB camera was used for image capture in a study conducted in narrow–row spaced (0.125 m) wheat infested with broadleaf weeds (Berge et al. 2008). Each image covered an area of 0.27 m^2. These researchers evaluated a prototype of an object–oriented image classification algorithm

(WeedFinder); plants were differentiated from bare soil based on the ratio between the green and red channel of the original color of each image. The contours of plant pixels in the binary image were extracted and weed crop discrimination was made based on size criteria. This prototype tended to underestimate, at some point, high actual weed densities and overestimate low actual weed densities. However, it should be mentioned that image and actual data were positively correlated. In a later study, Berge et al. (2012) used a novel machine vision called 'Weedcer' to analyze images captured with an RGB camera to estimate total ground coverage percentage of broadleaf weeds. Weed maps were created and used to provide recommendations for precision herbicide application. Herrera et al. (2014) proposed a strategy for discriminating between grass and broad–leaved weeds using a set of seven Hu invariant moments and six geometric shape descriptors (perimeter, diameter, minor axis length, major axis length, eccentricity and area). This combined strategy worked properly when the weeds were at an early growth stage, but according to authors, weed identification problems typically occur when the crop starts to overlap with weeds at later growth stages. To partially deal with this situation, weed density can be estimated at a sub–patch instead of patch level (Laursen et al. 2016). Under this concept, a coverage ratio vision algorithm (MoDiCoVi) was used to estimate ground coverage of grass and broadleaf weeds after processing images captured with a Basler Ace aca2000–50gc RGB camera. This approach was based on a close analysis of the contours of vegetation regions. By observing vegetation contours instead of individual plants, the problem of overlapping weed–crop leaves can be mitigated at some point. However, obtaining reliable results in crops like maize infested by grass weeds is a weakness of the method described.

In general, combining RGB imaging with the use of appropriate data processing algorithms is promising in the field of weed identification in horticultural crops. Recent research by Zheng et al. (2017) demonstrated 90.2–93.9% accuracy in classification between weeds and maize using RGB images and a SVM model (Zheng et al. 2017). There is evidence that the integration of RGB imaging along with machine learning algorithms can contribute to weed identification in more complex situations where weed species exhibit a strong botanical and morphological similarities with the crop (Kanatas et al. 2021a). Espejo-Garcia et al. (2020) demonstrated almost complete classification between crops and weed species exhibiting similar morphological characteristics with cultivated plants; tomato (*Solanum lycopersicum* L.) was differentiated from black nightshade (*Solanum nigrum* L.) while cotton was differentiated with velvetleaf (*Abutilon theophrasti* Medic.) after processing of RGB images (captured with a Nikon D700 camera) with a combination of CNN and SVM models. In another study, RGB image data were subjected to analysis using the CNN model with the k–FLBPCM algorithm resulting in up to 99.33% discrimination between wild radish (*Raphanus raphanistrum* L.) and canola (Le et al. 2020).

The Potential of Hyperspectral–Imaging Sensors for Weed Identification

Hyperspectral imaging combines the features of imaging and spectroscopy to collect spectral information over the full wavelength range for each pixel of the acquired image (Su et al. 2017). The data obtained provides images with a three–dimensional (3–D) structure containing one spectral dimension and two spatial dimensions (Andújar et al. 2016; Su 2020b). As a result, both spectral and image features of weeds and crops can be evaluated to improve the process of weed identification under real field conditions (Su 2020a).

Before describing scientific reports derived from field experiments, evidence on the potential of hyperspectral imaging techniques for weed identification is presented as derived from greenhouse trials. Fletcher and Turley (2017) used canopy hyperspectral narrowband data as input into the random forest model to discriminate between Palmer amaranth (*Amaranthus palmeri* S. Wats.] and cotton. Classification accuracy ranged between 77 and 89%. A Fieldspec 3 Full Range hyperspectral spectroradiometer was used to acquire hyperspectral images within the spectral range of 350 to 2500 nm. In another study conducted in pots under controlled greenhouse conditions, the same equipment was used to capture hyperspectral images (Fletcher and Reddy 2016) and the RF machine learning algorithm was again used for data post–processing. Recognition rates between Palmer amaranth, redroot pigweed (*Amaranthus retroflexus* L.), and soybean [*Glycine max* (L.) Merr.] were 93.8–100%. Gao et al. (2018) used a hyperspectral snapshot mosaic camera including two camera heads equipped with CMOSIS CMV2000 sensors and an embedded chip system to obtain hyperspectral data of weeds and maize grown together in a glasshouse. The authors used the RF model to classify maize, field bindweed (*Convolvulus arvensis* L.), Canada thistle [*Cirsium arvense* (L.) Scop.], and bitter dock (*Rumex obtusifolius* L.) with accuracies of 100, 78.9, 69.1 and 75.2%, respectively.

Another study by Zhang and Slaughter (2011) on six processing tomato cultivars used a CoolSNAP$_{cf}$ monochrome area camera equipped with a ImSpector V8 hyperspectral spectrograph to capture hyperspectral data of the tomato crop and two common summer weed species, redroot pigweed and black nightshade. Bayesian classification models were then used to discriminate crop plants and weeds with an average accuracy rate of 86.4%.

There are also promising reports derived from experiments carried out under open field conditions. Specifically, Zhang et al. (2019) scanned rice (*Oryza sativa* L.), barnyardgrass, and weedy rice (*Oryza sativa* L.) samples with a hyperspectral sensor and used a successive projection algorithm (SPA) for the selection of spectral features. Six wavelengths were selected (415 nm, 561 nm, 687 nm, 705 nm, 735 nm, and 1007 nm) within the spectral range between 380 nm to 1080 nm and were used to achieve 100, 100, and 92% correct classification rates for rice, barnyardgrass, and weedy rice respectively. In maize Pantazi et al. (2016) captured hyperspectral images were captured with an Inspector V9 line spectrograph which was combined with a monochromatic camera with a spectral range of 435–834 nm. MOG and SOM models were used for crop–weed discrimination and achieved recognition rates of 100% for maize and 94.44–98.15% for Canada thistle. In the same crop, Andújar et al. (2016) used a Kinect v2 device consisting of a depth camera, an RGB camera, and an infrared emitter. The results of the discriminant analysis that followed showed that all of the samples infested by broadleaf weeds (*Datura* spp., common cocklebur, common saltwort; *Salsola kali* L., and prostrate knotweed; *Polygonum aviculare* L.) were classified, as well as 92.3% of the samples infested with johnsongrass. In addition, there was a significant and strong ($R^2 = 0.70$) linear relationship between actual and hyperspectral data. In wheat, a Spectral Camera HS was used for line scanning (pushbroom sensor) to acquire hyperspectral images that were subjected to partial least squares discriminant analysis (PLS–DA). Wheat, grass weeds, broadleaf weeds, and bare soil were correctly classified with a precision of 85% (Herrmann et al. 2013). In processing tomato, another study was carried out early in this decade and provided interesting results. It is important that future research continues to focus on maximizing the accuracy of hyperspectral imaging practices for weed identification in order to deal with complex weed problems that mainly occur when the leaves of crop plants and weeds overlap at late growth stages (Su 2020b).

The Potential of Unmanned Aerial Vehicles (UAV) Mounted Sensors for Weed Identification

Unmanned aerial vehicles (UAVs) have become a common tool in precision horticulture and have potential to contribute to weed detection and identification in horticultural crops (Mohidem et al. 2021). UAV mounted sensors are a cost–effective method of collecting high resolution spatial crop data (Manfreda et al. 2018; Malamiri et al. 2021). Data are then processed with machine learning algorithms and subjected to classification techniques to create weed maps and implement site–specific weed management practices (Fernández-Quintanilla et al. 2018; Esposito et al. 2021; Mohidem et al. 2021). An example is the object–based image analysis (OBIA) methodology that identifies spatially and spectrally homogenous objects by grouping adjacent pixels according to a process called segmentation (López-Granados et al. 2016a, De Castro et al. 2018).

There are several recent scientific reports on the performance of UAV mounted sensors for weed discrimination in the field. For example, Gašparović et al. (2020) used a senseFly eBee UAV equipped with an integrated off–the–shelf senseFly S.O.D.A. RGB camera to create a weed map in oat fields (*Avena sativa* L.) where Roman chamomile (*Chamaemelum nobile* L.) and Canada thistle were the dominant weed species. UAV flying height was 110 m. Weeds and bare soil were classified according to the RF machine learning algorithm; recognition rates reached 89.0 and 87.1% for weeds and soil, respectively. Two consumer–grade RGB cameras (Canon

Powershot G15 and DJ Phantom 3 Professional) were mounted on a hexacopter MikroKopter UAV to obtain imaginary data of Canada thistle populations infesting wheat and barley fields (Rasmussen et al. 2019). UAV flights were carried out before crop harvest at the heights of 10, 20, 30, and 50 m. The use of a novel classifier called TopMaxExG used for visual threshold editing resulted in 92–97% correct identification of this noxious perennial species. UAV's have also been used for weed mapping in the sunflower (*Helianthus annuus* L.) crop (López-Granados et al. 2016a). RGB imagery was collected with two different cameras mounted separately in a quadrocopter MD4–1000 UAV. Flying heights were 30 and 60 m. The OBIA i.e., object–based–image analysis (OBIA) procedure was used to classify crop plants and weeds with accuracies of 100 and 85–100%, respectively. De Castro et al. (2018) used the same procedures and scanned three sunflower fields and two cotton fields to create weed maps. They achieved weed identification of 84.0 and 81.1% in cotton and sunflower, respectively. The MD4–1000 UAV model was used in another study conducted in maize to differentiate between maize and johnsongrass (López-Granados et al. 2016b). Flying heights were 30, 60, and 100 m. The UAV was equipped with an Olympus PEN E–PM1 RGB camera and a Tetracam mini–MCA–6 multispectral camera. The OBIA algorithm discriminated between maize plants and weeds achieving an accuracy rate of 86–92% at the flying height of 30 m. In wheat, an eBee fixed wing UAV equipped multispectral Canon S110 NIR camera was used to identify patches of milk thistle [*Silybum marianum* (L.) Gaertn.] in wheat fields where sterile oat was the dominant weed species. The maximum likelihood (ML) classifier was used to classify milk thistle and sterile oat with an accuracy of approximately 87% (Tamouridou et al. 2017).

However, there are some challenges associated with the use of UAVs. They are limited by lithium battery capacity and the UAV power weight ratio. The endurance time of flight is mostly limited to 30 min, and cannot exceed 60 min even under low load conditions. In addition, aerial imaging can be affected by the prevailing weather conditions. In contrast, ground–based weed detection systems are less dependent on weather conditions (Peteinatos et al. 2014). In addition, there is no doubt that UAV performance is promising on crops grown at wide row spacing. However, weed detection in crops grown at narrow row spacing such as winter cereals remains an obstacle (Torres-Sánchez et al. 2014). Moreover, another difficult task to accomplish using UAV images is to discriminate between different weed species when diverse weed communities are infesting a horticultural area (Fernández-Quintanilla et al. 2018). Future research should be focused on these issues to fully exploit the potential of UAV's for weed identification in horticultural fields.

5.3 Site-Specific Mechanical Weed Control

Mechanical operations have a leading role in the development of effective weed management practices, especially in organic horticulture (Bond and Grundy 2001). It should also be noted that it can be combined with other cultural practices to suppress weeds even in poorly competitive crops with very slow early growth rates (Alba et al. 2020; Kanatas and Gazoulis 2021). In general, mechanical weeding operations are widely used to control weeds in field crops grown for food, feed, industrial, and bioenergy production as well as in vegetables and several other high value crops (Pannacci and Tei 2014; Gazoulis et al. 2021; Nurk et al. 2017; Pannacci et al. 2017; Kanatas et al. 2021b). Between crop rows, rotary hoes, split–hoes, brush weeders, and spring tine harrows can be used for interrow cultivations while finger–weeders, torsion weeders, and flex–tine harrows are well suited for intrarow weed control (Pannacci et al. 2017; Hussain et al. 2018).

To maximize efficacy, mechanical operations should be conducted close to cultivated plants but should avoid mechanical damage to the crop (Rueda-Ayala et al. 2010; Hussain et al. 2018). Sensor–based weeders have the potential to differentiate between crops and weeds so as to ensure an optimal alignment of the machinery with the crop row and maximize the treated area of the field. Sensor–based mechanical weed control can be achieved with half autonomous tractor–pulled implements and autonomous robots. Half autonomous tractor–pulled implements are divided into machine vision systems, GNSS (global navigation satellite systems) sensor systems, and laser and ultrasonic sensors systems (Machleb et al. 2020).

To detect weed patches in the field, geo–referencing the crop plants via GNSS and storing their exact location in a plant map is the first option; an autonomous weeder can then work on weeds infesting the crop (Upadhyaya and Pérez-Ruiz 2012). As an alternative option, autonomous robots can be equipped with cameras to implement site–specific weed control treatments. Åstrand and Baerveldt (2002) describe a system including a forward–looking camera working with a grey level vision system and a near–infrared filter combined with a selective rotary hoe for weed control in sugar beet (*Beta vulgaris* L.). In greenhouse screening trials, the autonomous robot recognized all the sugar beet plants and the rotary hoe was able to effectively target the weed species. Van Evert et al. (2011) described another autonomous robot system equipped with a downward–looking camera for weed identification and a vertical rod weeder designed to cut weeds into pieces; bitter dock was the dominant weed species in this study conducted on a grassland. The success of weed detection was 92–93% and bitter dock regrowth after treatment was below 30% in two sites. Although these results are very promising, there is a need for further demonstration and validation under field conditions in a wider range of crops. If this technology is to be used commercially by farmers, they will require technical support with possible on–farm repairs that might be necessary to guarantee a continuous operating state of autonomous robots (Machleb et al. 2020).

There is more evidence on the use of tractor pulled machine vision systems guiding mechanical weed control treatments. Kunz et al. (2015) carried out a two–year field experiment with common lambsquarters, black bindweed (*Polygonum convolvulus* L.), prostrate knotweed, wild chamomile (*Matricaria chamomilla* L.), and purple deadnettle (*Lamium purpureum* L.) being the dominant weeds. A Kult–robocrop® hoe was used for precision inter–row hoeing alone or in combination with finger–weeders for intrarow weed control. Operations were guided from a camera and resulted in 87 and 89% lower weed density in sugar beet and soybean, respectively. In their next field trials in sugar beet, these researchers obtained a reduction of 82% in weed density after implementing mechanical weed control with the use of automatic steering technologies (Kunz et al. 2016). Similar results were seen in eight two–year field trials carried out in maize, soybean and sugar beet fields infested mainly with common lambsquarters, black bindweed, and field penny–cress (*Thlaspi arvense* L.); camera steered mechanical weed control resulted in 78% weed control efficacy and improved the yield of all three studied crops (Kunz et al. 2018).

In high value crops, Fennimore et al. (2014) evaluated a smart, rotating cultivator (Robocrop) that utilized machine–vision guidance to align a rotating disk with the crop plants to control weeds and remove undesired crop plants in Chinese cabbage [*Brassica rapa* L. subsp. *chinensis* (L.) Hanelt], celery (*Apium graveolens* L.), lettuce, and radicchio (*Cichorium intybus* L. var. *foliosum*). Weed density reductions of 85% were achieved demonstrating that such systems can provide good weed control in transplanted vegetables. Successful intrarow weed control was also achieved in two field trials conducted in transplanted onion (*Alium cepa* L.) and white cabbage using an intelligent, camera–steered Robovator intra–row weeder (Melander et al. 2015). This system includes a machine vision system for single crop plant identification and a pair of tines that work on each crop row. The tines open when a crop plant is encountered and close after passing the crop plant. In this way, weeds are removed mechanically from the intrarow space. In any case, it can be noted that weed identification is easier in such situations where cultivated plants are larger than weeds. As a result, camera–steered mechanical treatments tend to be highly effective. However, the same equipment was tested in direct–seeded broccoli (*Brassica oleracea* L. var. *italica*) and transplanted lettuce; and promising results were obtained also for the direct–seeded crop (Lati et al. 2016). In two out of a total of three field experiments, intelligent weeding reduced weed density by 52–70% compared to conventional practices. The potential of distance (LiDAR and ultrasonic) sensors for weed identification have been already presented in previous sections of the chapter. However, there are not many reports on their performance for guiding mechanical weed control operations. A rare example is found in the study by Rueda-Ayala et al. (2015). These authors used ultrasonic readings to guide the function of a spring tine harrow. Weed control reached up to 91% in some cases without causing crop injury.

In perennial orchards and vineyards where soil tillage can be harmful for tree and vine roots, mowing is frequently used to control weeds (Travlos et al. 2018; Mia et al. 2020). There are some recent reports of weed control in perennial crops by means of precision horticulture methodologies and particularly with the use of autonomous mowers. Hossain and Komatsuzaki (2021) used a Husqvarna RLM automower 430X, navigated by GPS and with a working capacity of 3200 m^2 day^{-1} (± 20%) and a mowing height of 6 cm. The experiment was carried out in a Japanese pear orchard [*Pyrus pyrifolia* (Burm.) Nak.]. Although operation stopped when the mower hit fallen fruits and physical obstacles, this study showed that the autonomous mower can outperform conventional mowers in small acreages. Sportelli et al. (2021) described another study in a vineyard where a Husqvarna AWD automower 535 with working capacity of 3500 (± 20%) m^2 day^{-1} and mowing height of 6 cm was evaluated; weed canopy height and total weed biomass were lower in plots mown with the autonomous machine compared to conventional weed management practices.

5.4 Site-Specific Herbicide Application

One of the most important targets of precision horticulture methodologies and technological applications developed for weed management is to ensure that no more herbicide is used than needed (Rydahl 2003; Hamouz et al. 2013; Xu et al. 2018). Sensor-based, precise herbicide applications can help reduce herbicide inputs in horticultural systems and, thus the environmental risks associated with herbicide use, while maintaining high yields and producing high-quality products (López-Granados 2011; Dammer 2016; Chantre and González-Andújar 2020). According to Christensen et al. (2009), the basic parts of site-specific herbicide applications include a reliable weed identification system, a model to process data through the use of the appropriate decision-making algorithms, and precision weed control implements such as sprayers equipped with a series of controllable nozzles that enable spatially variable applications of herbicides. During the last two decades, there have been a number of scientific studies which have quantified herbicide savings obtained by adopting precision horticulture methodologies for weed management. Site-specific applications result in optimal herbicide use; efficacy is maintained without applying higher amounts of agrochemicals than those that are actually required to achieve sufficient levels of weed control. It should be remembered that any strategy which aims to optimize herbicide use has unique importance for Europe in light of the targets of the European Green Deal to reduce pesticide inputs in horticulture by at least 50%.

In early studies, Tian (2002) developed and evaluated a sensor-based precision herbicide application system; a four-level application-rate scheme was selected to spray weed patches that were identified. Individual spray nozzles of a prototype sprayer were controlled separately. Tian (2002) achieved herbicide savings of 6–18% and 52–71% in plots where intermediate and high weed density was observed, respectively. In maize, Yang et al. (2003) captured weed coverage and patchiness data through RGB-imaging and then used a fuzzy-logic algorithm to process imaging data and create a weed map for site-specific herbicide application. Their simulations indicated that significant herbicide savings could be achieved by using this approach. Gerhards and Christensen (2003) used three bi-spectral cameras to identify weed patches in sugar beet, maize, and winter cereal fields. Data processing was carried out using a decision algorithm for patch spraying (DAPS). Averaged over four experimental years, targeted herbicide use reduced herbicide inputs up to 41%, 78% and 90% in sugar beet, maize, and winter cereal fields, respectively, compared to conventional weed management practices. In a three-year field trial conducted in 13 fields cultivated with cereals or peas, sensor-based herbicide application resulted in 22.8 and 27.9% herbicide savings, respectively

(Dammer and Wartenberg 2007). Similar results were observed in a field study by Gerhards and Oebel (2006) carried out in winter oilseed rape, maize, winter wheat, spring barley, and sugar beet. The authors collected and processed imaging data to map weed presence and perform site-specific herbicide applications. The efficacy of weed control ranged between 85% to 98%, while herbicide savings for the control of grass and broadleaf weeds were 20–79 and 6–81%, respectively. These precision horticulture methodologies can contribute to the development of integrated weed management (IWM) strategies, and this was demonstrated by Wiltshire et al. (2003) who combined precision mechanical weed control operations with banded herbicide applications and reduced herbicide inputs by 70% in sugar beet.

More recent studies have continued to demonstrate that precise herbicide applications can lead to sufficient levels of weed control and significant herbicide savings. In winter wheat fields infested mainly with catchweed bedstraw, Canada thistle, couch grass [*Elytrigia repens* (L.) Desv. ex Nevski.] and false mayweed [*Tripleurospermum inodorum* (L.) Sch.Bip.], Hamouz et al. (2013) created herbicide maps and used a real time kinematic (RTK) navigation system to target herbicide application. The authors measured herbicide savings of up to 78, 84, and 91% for clopyralid, metsulfuron methyl plus tribenuron-methyl, and pinoxaden, respectively. In carrots, Dammer (2016) demonstrated a reduction in herbicide inputs of 30–43% compared to standard herbicide use practices. The authors used image data collected with a camera sensor and processed using the 'Weeds' software. NDVI values were used as indicators to set thresholds for weed control and spraying was conducted with a BBG 4000 field sprayer equipped with a 'VarioSelect' system including multiple nozzle bodies. Weed coverage level was converted to a voltage between 1 V to 4 V that served as a signal for the sprayer to operate. More recently, Xu et al. (2018) evaluated an automated weed mapping and variable-rate herbicide spraying (VRHS) system for row crops resulting in optimal herbicide use. The system included a lateral histogram-based algorithm for rapid weed mapping using image data collected with a MV-EM040M industrial grayscale camera. Power et al. (2013) assessed the effectiveness of automated spot-treatments and found significantly lower levels of injury to non-target plants compared to conventional practices. In a study by Gonzalez-de-Soto et al. (2016), a robotized patch sprayer consisting of an autonomous mobile robot based on a commercial agricultural vehicle chassis and a direct-injection spraying boom was used with on-board and remote sensors to collect weed data and target herbicide applications. Field trials showed that the system was reliable and successfully treated over 99% of weeds that were correctly identified. Partel et al. (2019) set up two different experimental scenarios to simulate a vegetable field and assess the performance of an intelligent sprayer. The intelligent sprayer included individual nozzle control, a low-cost pump, a real time kinematic GPS (RTK-GPS), three video cameras, a smart controller to control sensors and actuators, and a computational unit for image processing and weed detection. This system achieved a spraying accuracy of 78% and 91% in cases where applications were performed on real and artificial weed seedlings, respectively.

There has also been significant research effort in the development of computer-based Decision Support Systems (DSS) for precision weed management (PWM);

DSSs integrate decision algorithms quantifying the special requirements for weed control in a particular field (Chantre and González-Andújar 2020; Kanatas et al. 2020). One of the first DSSs developed for weed management was the Danish Crop Protection Online (CPO) that was introduced by Rydahl (2003). Its recommendations have been shown to result in herbicide savings of 35, 38, 44 and 80% in spring cereals, fodder beet (*Beta vulgaris* L. subsp. *vulgaris* var. *crassa*), winter cereals, and sugar beet respectively, compared to reference herbicide treatments. The performance of 'CPOWeeds', a version of 'CPO' adjusted to the Spanish conditions, was tested in winter cereal field trials carried out over three experimental years by Montull et al. (2014). These authors obtained satisfactory efficacy for herbicides applied according to 'CPOWeeds' recommendations and herbicide inputs were reduced by up to 30% in 9 out of their 17 experiments. According to Sønderskov et al. (2015), the recommendations of another DSS based on the Danish 'CPO', DSSHerbicide, resulted in 20–40% lower herbicide inputs in field trials conducted in winter wheat fields. AvenaNET and ValllicoNET are examples of other web-based DSSs that have been recently developed for the management of rigid ryegrass (*Lolium rigidum* Gaud.) and sterile oat, respectively, in Spain. The PC versions of these DSSs, namely LOLIUM-PC and AVENA-PC, have been reported to reduce herbicide inputs for the control of rigid ryegrass and sterile oat by 57 and 65%, respectively, in comparison to the conventional practices (González-Andújar 2020).

Except for their great potential in reducing herbicide inputs, remote sensing and precision horticulture methodologies are very useful for the evaluation of herbicide performance under real field conditions. There is evidence that weed responses to stress factors can be estimated through sensor-based techniques (Peteinatos et al. 2016). Herbicide applications induce stress in treated plants. Carotenoid inhibition, leaf bleaching, and lower chlorophyll content in the meristematic tissues are the typical outcomes of herbicide application affecting plant health (Peñuelas and Filella 1998). Vegetation indices such as NDVI can detect changes in vegetation growth caused by herbicide application. Henry et al. (2004) stated that the process of red light's absorption followed by energy reflectance in the NIR region is an attribute of healthy vegetation. Given the way NDVI is computed (Tremblay et al. 2009), a decrease in the health status of vegetation can be detected from reduced NDVI values. Novel rapid methods have been recently developed for the *in-situ* evaluation of herbicide efficacy by using NDVI measurements as an indicator of herbicide performance (Kanatas et al. 2021c; Travlos et al. 2021). In both of the studies mentioned, NDVI was measured with a 'GreenSeeker' handheld device in a short time after treatment. Positive and strong correlations were obtained between NDVI and weed biomass highlighting the reliability of sensor-based methods. In addition, such methods tend to be more reliable than visual ratings and less time-consuming than destructive measurements such as weed biomass evaluations (Lewis et al. 2014; Streibig et al. 2014).

Another approach is the estimation of the maximum quantum yield of PS II i.e., the ratio of variable to maximum fluorescence (F_v/F_m), as an indicator of herbicide efficacy. Hassannejad et al. (2020) used a 'Handy-PEA' portable fluorometer to measure this ratio and evaluate the efficacies of bentazon and nicosulfuron on common cocklebur populations. These authors observed the high efficacy of bentazon and the low efficacy of nicosulfuron at just 24 h after treatment. Kanatas et al. (2021c) denoted that this particular ratio was reliable for screening herbicides that were effective on Palmer amaranth populations collected from maize, cotton, and grain sorghum fields in Greece. These authors obtained strong and significant correlations between the F_v/F_m ratio and weed biomass, weed height, and weed canopy cover. 'FluorPen FP 110' was the portable fluorometer used for evaluations.

Moreover, sensor-based methods can contribute to the rapid detection of weed populations that are potentially resistant to herbicides. In the study by Zhang et al. (2016), barnyardgrass was the weed species under study. After treatments with a PS II inhibitor (i.e. propanil), resistant and susceptible populations were successfully discriminated by measuring the maximum quantum yield of PS II. Even after treatments with Acetyl-CoA-carboxylase (ACCase) or Acetolactate Synthase (ALS) inhibitors, the leaf chlorophyll fluorescence assay detected resistant populations at 8 days after treatment. This period is much shorter than the 21 days that are required to confirm resistance with conventional whole-plant assays. In the study by Kaiser et al. (2013), accessions of black grass were cultivated in tissue culture plates containing nutrient agar and different rates of fenoxaprop-*p*-ethyl and mesosulfuron plus iodosulfuron. These researchers measured the maximum quantum efficiency of the PSII 3 h after transplanting and then for 7 days every 24 h. Chlorophyll fluorescence imaging corresponded well with the standard whole-plant pot tests in the glasshouse but was significantly less time consuming. Concerning the potential of sensor-based methods to detect the presence of herbicide-resistant weeds in the field, an example is found in the study by Travlos et al. (2021). These authors measured NDVI above weed canopy and between crop rows at two weeks after treatment, by using a handheld optoelectronic sensor. In several winter wheat fields, weed canopies were characterized from high NDVI values. Weed seeds were collected from these sites to perform greenhouse dose-response experiments. It was found that populations exhibiting high NDVI values in the field were either resistant or were developing resistance to ACCase- and ALS-inhibiting herbicides. Research should continue to further optimize such methods in more crops under real field conditions so as to provide with farmers actual recommendations for their field activities regarding weed management.

5.5 Concluding Remarks

Intensive research has been carried out over the past 15–20 years to develop intelligent and reliable precision horticulture methodologies and a number of studies have demonstrated successful precision identification and treatment of weeds. However, the adoption of site-specific weed management methods by farmers and growers remains sporadic (Fountas et al. 2005; Griffin 2016; Fernández-Quintanilla et al. 2018). An example can be found in studies reporting on the results from surveys in Denmark and Midwest USA (Fountas et al. 2005). In Denmark, only 28 and 32% of farmers and growers were willing to use weed mapping and precise herbicide applications, respectively, while the corresponding percentages for USA were both below 10%. More recent surveys of precision practices used in USA revealed that farmers and growers did not include weed mapping as one of the technologies likely to be adopted (Lambert et al. 2015; Griffin 2016).

In the study of Kernecker et al. (2020), some practitioners highlighted that the majority of farmers were not familiar with the smart technologies that were available. In fact, most farmers are frustrated by the rate of change in technology, and subsequently avoid taking the risk to rely on the performance of modern high-tech tools (Colas et al. 2020; Kanatas et al. 2020). To mitigate these problems, further development work should be done in collaboration with stakeholders to ensure the technology addresses their needs and demands (Ingram and Gaskell 2019). It is vital to carry out field surveys, host workshops and group meetings in order to interact with farmers and help them understand precision horticulture methodologies for weed management (Colas et al. 2020).

There are also other factors that pose obstacles to increasing the adoption of smart weed management systems. Although the progress is obvious, the challenge to obtain reliable results under the real field conditions and deal with the real needs of farmers remains a challenge. One of the most common problems reported for weed detection was crop-weed overlapping at advanced growth stages. Other studies have reported classification problems between grass weeds and monocotyledonous crops or between broadleaf weeds and dicotyledonous crops. Future research should continue to focus on developing systems that can mitigate such problems and provide farmers with accurate results. As an example, a novel strategy proposed by Su (2020b) includes crop plant signaling as a new robot-plant interaction technique that allows weed-crop differentiation based on photographing of a machine-readable fluorescence signal applied to crop plants rather than the weed.

In addition, farmers should be provided with information on the economic benefits associated with the use of precision horticulture methodologies for site specific weed management (Foley et al. 2011; Van Evert et al. 2017). There may be benefit in giving economic incentives to the farmers in order to encourage them to use new technologies instead of conventional agronomic practices (Lati et al. 2021). A crucial issue is also to persuade farmers about herbicide savings that can be achieved through the adoption of site-specific weed management while improving crop productivity. In other words, farmers need to know that shifting to more environmentally friendly weed control options is not an obstacle for achieving high yields (Kanatas et al. 2020).

Conclusively, precision horticulture and site–specific weed management (SSWM) practices based on remote sensing techniques can contribute to the development of smart, eco-friendly weed management strategies. Weed identification is a challenging task to accomplish under field conditions. Non–imaging and imaging sensors can be used for this purpose. Once data are captured, they can be processed with machine learning algorithms so as to classify between either weeds and bare soil or between weeds and crop plants. Data can be used to create weed maps so as to perform precise mechanical weed control operations and site-specific herbicide applications. Precision horticulture methodologies have been shown to result in significant herbicide savings. However, the use of site-specific weed management tools by the farmers remains low. To increase adoption, experts need to establish stronger relationships with the growers to familiarize them with the concepts of precision horticulture and site-specific weed management.

Competing Interests The authors have no conflicts of interest to declare that are relevant to the content of this chapter.

References

Akbarzadeh S, Paap A, Ahderom S, Apopei B, Alameh K (2018) Plant discrimination by support vector machine classifier based on spectral reflectance. Comput Electron Agric 148:250–258

Alba OS, Syrovy LD, Duddu HS, Shirtliffe SJ (2020) Increased seeding rate and multiple methods of mechanical weed control reduce weed biomass in a poorly competitive organic crop. Field Crop Res 245:107648

Andújar D, Escola A, Dorado J, Fernández-Quintanilla C (2011a) Weed discrimination using ultrasonic sensors. Weed Res 51:543–547

Andújar D, Ribeiro Á, Fernández-Quintanilla C, Dorado J (2011b) Reliability of a visual recognition system for detection of johnsongrass (*Sorghum halepense*) in corn. Weed Technol 25:645–651

Andújar D, Weis M, Gerhards R (2012) An ultrasonic system for weed detection in cereal crops. Sensors 12:17343–17357

Andújar D, Escolà A, Rosell-Polo JR, Fernández-Quintanilla C, Dorado J (2013a) Potential of a terrestrial LiDAR–based system to characterize weed vegetation in maize crops. Comput Electron Agric 92:11–15

Andújar D, Rueda-Ayala V, Moreno H, Rosell-Polo JR, Valero C, Gerhards R, Fernández-Quintanilla C, Dorado J, Griepentrog HW (2013b) Discriminating crop, weeds and soil surface with a terrestrial LIDAR sensor. Sensors 13:14662–14675

Andújar D, Dorado J, Fernández-Quintanilla C, Ribeiro A (2016) An approach to the use of depth cameras for weed volume estimation. Sensors 16:972

Åstrand B, Baerveldt AJ (2002) An agricultural mobile robot with vision–based perception for mechanical weed control. Auton Robot 13:21–35

Berge TW, Aastveit AH, Fykse H (2008) Evaluation of an algorithm for automatic detection of broad-leaved weeds in spring cereals. Precis Agric 9:391–405

Berge TW, Goldberg S, Kaspersen K, Netland J (2012) Towards machine vision based site–specific weed management in cereals. Comput Electron Agric 81:79–86

Bond W, Grundy AC (2001) Non-chemical weed management in organic farming systems. Weed Res 41:383–405

Bronson KF, French AN, Conley MM, Barnes EM (2021) Use of an ultrasonic sensor for plant height estimation in irrigated cotton. Agron J 113:2175–2183

Brown RB, Noble SD (2005) Site–specific weed management: sensing requirements—what do we need to see? Weed Sci 53:252–258

Burgos-Artizzu XP, Ribeiro A, Tellaeche A, Pajares G, Fernández-Quintanilla C (2009) Improving weed pressure assessment using digital images from an experience–based reasoning approach. Comput Electron Agric 65:176–185

Cerovic ZG, Samson G, Morales F, Tremblay N, Moya I (1999) Ultraviolet–induced fluorescence for plant monitoring: present state and prospects. Agronomie 19:543–578

Chantre GR, González-Andújar JL (2020) Decision support systems for weed management. Springer, Cham
Chauhan BS (2020) Grand challenges in weed management. Front Agron 1:3
Christensen S, Søgaard HT, Kudsk P, Nørremark M, Lund I, Nadimi ES, Jørgensen R (2009) Site-specific weed control technologies. Weed Res 49:233–241
Colaço AF, Schaefer M, Bramley RG (2021) Broadacre mapping of wheat biomass using ground–based LiDAR technology. Remote Sens 13:3218
Colas F, Cordeau S, Granger S, Jeuffroy M-H, Pointurier O, Queyrel W, Rodriguez A, Villerd J, Colbach N (2020) Co-development of a decision support system for integrated weed management: contribution from future users. Eur J Agron 114:126010
Cox C (2006) Ten reasons not to use pesticides. J Pestic Ref 26:10–12
Dammer KH (2016) Real–time variable–rate herbicide application for weed control in carrots. Weed Res 56:237–246
Dammer KH, Wartenberg G (2007) Sensor–based weed detection and application of variable herbicide rates in real time. Crop Prot 26:270–277
De Castro AI, Torres-Sánchez J, Peña JM, Jiménez-Brenes FM, Csillik O, López-Granados F (2018) An automatic random forest–OBIA algorithm for early weed mapping between and within crop rows using UAV imagery. Remote Sens 10:285
De Souza MF, do Amaral LR, de Medeiros Oliveira SR, Coutinho MAN, Netto CF (2020) Spectral differentiation of sugarcane from weeds. Biosyst Eng 190:41–46
Deng W, Huang Y, Zhao C, Chen L, Wang X (2016) Bayesian discriminant analysis of plant leaf hyperspectral reflectance for identification of weeds from cabbages. Afr J Agric Res 11:551–562
Ehlert D, Adamek R, Horn HJ (2009) Laser rangefinder–based measuring of crop biomass under field conditions. Precis Agric 10:395–408
El-Naggar AG, Jolly B, Hedley CB, Horne D, Roudier P, Clothier BE (2021) The use of terrestrial LiDAR to monitor crop growth and account for within–field variability of crop coefficients and water use. Comput Electron Agric 190:106416
Espejo-Garcia B, Mylonas N, Athanasakos L, Fountas S, Vasilakoglou I (2020) Towards weeds identification assistance through transfer learning. Comput Electron Agric 171:105306
Esposito M, Crimaldi M, Cirillo V, Sarghini F, Maggio A (2021) Drone and sensor technology for sustainable weed management: a review. Chem Biol Technol Agric 8:1–11
Farooque AA, Chang YK, Zaman QU, Groulx D, Schumann AW, Esau TJ (2013) Performance evaluation of multiple ground based sensors mounted on a commercial wild blueberry harvester to sense plant height, fruit yield and topographic features in real–time. Comput Electron Agric 91:135–144
Felton WL, McCloy KR (1992) Spot spraying. Agric Eng 73:9–12
Fennimore SA, Smith RF, Tourte L, LeStrange M, Rachuy JS (2014) Evaluation and economics of a rotating cultivator in bok choy, celery, lettuce, and radicchio. Weed Technol 28:176–188
Fernández-Quintanilla C, Peña JM, Andújar D, Dorado J, Ribeiro A, López-Granados F (2018) Is the current state of the art of weed monitoring suitable for site-specific weed management in arable crops? Weed Res 58:259–272
Fletcher RS, Reddy KN (2016) Random forest and leaf multispectral reflectance data to differentiate three soybean varieties from two pigweeds. Comput Electron Agric 128:199–206
Fletcher RS, Turley RB (2017) Employing canopy hyperspectral narrowband data and random forest algorithm to differentiate palmer amaranth from colored cotton. Am J Plant Sci 8:3258
Foley JA, Ramankutty N, Brauman KA, Cassidy ES, Gerber JS, Johnston M et al (2011) Solutions for a cultivated planet. Nature 478:337–342
Fountas S, Blackmore S, Ess D, Hawkins S, Blumhoff G, Lowenberg-Deboer J, Sorensen CG (2005) Farmer experience with precision agriculture in Denmark and the US Eastern Corn Belt. Precis Agric 6:121–141
Gai J, Tang L, Steward BL (2020) Automated crop plant detection based on the fusion of color and depth images for robotic weed control. J Field Robot 37:35–52
Gao J, Nuyttens D, Lootens P, He Y, Pieters JG (2018) Recognising weeds in a maize crop using a random forest machine–learning algorithm and near–infrared snapshot mosaic hyperspectral imagery. Biosyst Eng 170:39–50

Gašparović M, Zrinjski M, Barković Đ, Radočaj D (2020) An automatic method for weed mapping in oat fields based on UAV imagery. Comput Electron Agric 173:10538

Gazoulis I, Kanatas P, Papastylianou P, Tataridas A, Alexopoulou E, Travlos I (2021) Weed management practices to improve establishment of selected lignocellulosic crops. Energies 14:2478

Gerhards R, Christensen S (2003) Real-time weed detection, decision making and patch spraying in maize, sugarbeet, winter wheat and winter barley. Weed Res 43:385–392

Gerhards R, Oebel H (2006) Practical experiences with a system for site-specific weed control in arable crops using real-time image analysis and GPS-controlled patch spraying. Weed Res 46: 185–193

Gerhards R, Weis M, Gutjahr C, Schulz J, Jancker H (2012) Untersuchungen zur automatisierten Unkrauterkennung in Kulturpflanzenbeständen mit Hilfe des Sensorsystems MiniVeg. In: Abstracts of the 17th Workshop Computer–Bildanalyse in der Landwirtschaft, Vol. 78 of Bornimer Agrartechnische Berichte. ATB—Leibniz–Institut für Agrartechnik Potsdam–Bornim e.V. Ihinger Hof, Potsdam–Bornim and Renningen, 5–9 May 2012

Gharde Y, Singh PK, Dubey RP, Gupta PK (2018) Assessment of yield and economic losses in agriculture due to weeds in India. Crop Prot 107:12–18

Gianessi LP (2013) The increasing importance of herbicides in worldwide crop production. Pest Manag Sci 69:1099–1105

González-Andújar JL (2020) AvenaNET and VallicoNET: DSS for *Avena sterilis* and *Lolium rigidum* control in Spanish dryland cereal crops. In: Chantre G, González-Andújar J (eds) Decision support systems for weed management. Springer, Cham, pp 299–309

Gonzalez-de-Soto M, Emmi L, Perez-Ruiz M, Aguera J, Gonzalez-de-Santos P (2016) Autonomous systems for precise spraying–evaluation of a robotised patch sprayer. Biosyst Eng 146:165–182

Griffin TW (2016) Adoption of precision agricultural technology in Kansas. https://www.agmanager.info/adoption-precision-agriural-technology-kansas. Accessed 31 Oct 2021

Hamouz P, Hamouzová K, Holec J, Tyšer L (2013) Impact of site-specific weed management on herbicide savings and winter wheat yield. Plant Soil Environ 59:101–107

Hassannejad S, Ghafarbi SP, Lotfi R (2020) The effect of nicosulfuron and bentazon on photosynthetic performance of common cocklebur (*Xanthium strumarium* L.). Environ Sustainability Indic 6:100026

Heap I (2020) International survey of herbicide resistant weeds. http://www.weedscience.org/Pages/RecentCases.aspx. Accessed 29 Oct 2021

Henry WB, Shaw DR, Reddy KR, Bruce LM, Tamhankar HD (2004) Remote sensing to distinguish soybean from weeds after herbicide application. Weed Technol 18:594–604

Herrera PJ, Dorado J, Ribeiro A (2014) A novel approach for weed type classification based on shape descriptors and a fuzzy decision–making method. Sensors 14:15304–15324

Herrmann I, Shapira U, Kinast S, Karnieli A, Bonfil D (2013) Ground–level hyperspectral imagery for detecting weeds in wheat fields. Precis Agric 14:637–659

Hossain MZ, Komatsuzaki M (2021) Weed management and economic analysis of a robotic lawnmower: a case study in a Japanese pear orchard. Agriculture 11:113

Hussain M, Farooq S, Merfield C, Jabran K (2018) Mechanical weed control. In: Jabran K, Chauhan B (eds) Nonchemical weed control. Academic Press, Cambridge, MA, USA, pp 133–155

Ingram J, Gaskell P (2019) Searching for meaning: co-constructing ontologies with stakeholders for smarter search engines in agriculture. NJAS Wageningen J Life Sci 90-91:100300

Kaiser YI, Menegat A, Gerhards R (2013) Chlorophyll fluorescence imaging: a new method for rapid detection of herbicide resistance in *Alopecurus myosuroides*. Weed Res 53:399–406

Kanatas PJ, Gazoulis I (2021) The integration of increased seeding rates, mechanical weed control and herbicide application for weed management in chickpea (*Cicer arietinum* L.). Phytoparasitica 50:255

Kanatas P, Travlos IS, Gazoulis I, Tataridas A, Tsekoura A, Antonopoulos N (2020) Benefits and limitations of decision support systems (DSS) with a special emphasis on weeds. Agronomy 10: 548

Kanatas PJ, Travlos IS, Gazoulis J, Antonopoulos N, Tsekoura A, Tataridas A, Zannopoulos S (2020a) The combined effects of false seedbed technique, post–emergence chemical control and cultivar on weed management and yield of barley in Greece. Phytoparasitica 48:131–143

Kanatas P, Travlos I, Papastylianou P, Gazoulis I, Kakabouki I, Tsekoura A (2020b) Yield, quality and weed control in soybean crop as affected by several cultural and weed management practices. Not Bot Hort Agrobot Cluj–Napoca 48:329–341

Kanatas P, Gazoulis I, Zannopoulos S, Tataridas A, Tsekoura A, Antonopoulos N, Travlos I (2021a) Shattercane (*Sorghum bicolor* (L.) Moench Subsp. *Drummondii*) and weedy sunflower (*Helianthus annuus* L.)—crop wild relatives (CWRs) as weeds in agriculture. Diversity 13:463

Kanatas P, Antonopoulos N, Gazoulis I, Travlos IS (2021b) Screening glyphosate–alternative weed control options in important perennial crops. Weed Sci 69:704

Kanatas P, Tataridas A, Dellaportas V, Travlos I (2021c) First report of *Amaranthus palmeri* S. Wats. in cotton, maize and sorghum in Greece and problems with its management. Agronomy 11:1721

Kavdr I (2004) Discrimination of sunflower, weed and soil by artificial neural networks. Comput Electron Agric 44:153–160

Kernecker M, Knierim A, Wurbs A, Kraus T, Borges F (2020) Experience versus expectation: farmers' perceptions of smart farming technologies for cropping systems across Europe. Precis Agric 21:34–50

Kokalj M, Prikeržnik M, Kreft S (2017) FTIR spectroscopy as a tool to detect contamination of rocket (*Eruca sativa* and *Diplotaxis tenuifolia*) salad with common groundsel (*Senecio vulgaris*) leaves. J Sci Food Agric 97:2238–2244

Krause GH, Weis E (1991) Chlorophyll fluorescence and photosynthesis: the basics. Annu Rev Plant Physiol Plant Mol Biol 42:313–349

Kunz C, Weber JF, Gerhards R (2015) Benefits of precision farming technologies for mechanical weed control in soybean and sugar beet—comparison of precision hoeing with conventional mechanical weed control. Agronomy 5:130–142

Kunz C, Weber JF, Gerhards R (2016) Comparison of different mechanical weed control strategies in sugar beets. Jul Kühn Archiv 452:446

Kunz C, Weber JF, Peteinatos GG, Sökefeld M, Gerhards R (2018) Camera steered mechanical weed control in sugar beet, maize and soybean. Precis Agric 19:708–720

Lambert DM, Paudel KP, Larson JA (2015) Bundled adoption of precision agriculture technologies by cotton producers. J Agric Resour Econ 40:325–345

Lati RN, Siemens MC, Rachuy JS, Fennimore SA (2016) Intrarow weed removal in broccoli and transplanted lettuce with an intelligent cultivator. Weed Technol 30:655–663

Lati RN, Rasmussen J, Andujar D, Dorado J, Berge TW, Wellhausen C, Pflanz M, Nordmeyer H, Schirrmann M, Eizenberg H, Neve P, Jørgensen RN, Christensen S (2021) Site-specific weed management—constraints and opportunities for the weed research community: insights from a workshop. Weed Res 61:147–153

Laursen MS, Jørgensen RN, Midtiby HS, Jensen K, Christiansen MP, Giselsson TM, Mortensen AK, Jensen PK (2016) Dicotyledon weed quantification algorithm for selective herbicide application in maize crops. Sensors 16:1848

Le VNT, Ahderom S, Alameh K (2020) Performances of the LBP based algorithm over CNN models for detecting crops and weeds with similar morphologies. Sensors 20:2193

Legg M, Bradley S (2020) Ultrasonic arrays for remote sensing of pasture biomass. Remote Sens 12:111

Lewis DF, Jeffries MD, Gannon TW, Richardson RJ, Yelverton FH (2014) Persistence and bioavailability of aminocyclopyrachlor and clopyralid in turfgrass clippings: recycling clippings for additional weed control. Weed Sci 62:493–500

Li Y, Sun Z, Zhuang X, Xu L, Chen S, Li M (2003) Research progress on microbial herbicides. Crop Prot 22:247–252

Liu B, Bruch R (2020) Weed detection for selective spraying: a review. Curr Robot Rep 1:19–26

Longchamps L, Panneton B, Samson G, Leroux GD, Thériault R (2010) Discrimination of corn, grasses and dicot weeds by their UV–induced fluorescence spectral signature. Precis Agric 11: 181–197

López-Granados F (2011) Weed detection for site-specific weed management: mapping and real-time approaches. Weed Res 51:1–11

López-Granados F, Torres-Sánchez J, Serrano-Pérez A, de Castro AI, Mesas-Carrascosa FJ, Pena JM (2016a) Early season weed mapping in sunflower using UAV technology: variability of herbicide treatment maps against weed thresholds. Precis Agric 17:183–199

López-Granados F, Torres-Sánchez J, De Castro AI, Serrano-Pérez A, Mesas-Carrascosa FJ, Peña JM (2016b) Object–based early monitoring of a grass weed in a grass crop using high resolution UAV imagery. Agron Sustain Dev 36:1–12

Machleb J, Peteinatos GG, Kollenda BL, Andújar D, Gerhards R (2020) Sensor–based mechanical weed control: present state and prospects. Comput Electron Agric 176:105638

Malamiri HRG, Aliabad FA, Shojaei S, Morad M, Band SS (2021) A study on the use of UAV images to improve the separation accuracy of agricultural land areas. Int J Remote Sens 184: 106079

Manfreda S, McCabe ME, Miller PE, Lucas R, Madrigal VP, Mallinis G, Ben Dor E, Helman D, Estes L, Ciraolo G, Müllerová J, Tauro F, De Lima MI, De Lima JLMP, Maltese A, Frances F, Caylor K, Kohv M, Perks M, Ruiz-Pérez G, Su Z, Vico G, Toth B (2018) On the use of unmanned aerial systems for environmental monitoring. Remote Sens 10:641

Marchant JA, Onyango CM (2003) Comparison of a Bayesian classifier with a multilayer feed–forward neural network using the example of plant/weed/soil discrimination. Comput Electron Agric 39:3–22

Melander B, Lattanzi B, Pannacci E (2015) Intelligent versus non–intelligent mechanical intra–row weed control in transplanted onion and cabbage. Crop Prot 72:1–8

Mia MJ, Massetani F, Murri G, Facchi J, Monaci E, Amadio L, Neri D (2020) Integrated weed management in high density fruit orchards. Agronomy 10:1492

Mohidem NA, Che'Ya NN, Juraimi AS, Fazlil Ilahi WF, Mohd Roslim MH, Sulaiman N, Mohd Noor N (2021) How can unmanned aerial vehicles be used for detecting weeds in agricultural fields? Agriculture 11:1004

Montull J, Sønderskov M, Rydahl P, Bøjer O, Taberner A (2014) Four years validation of decision support optimising herbicide dose in cereals under Spanish conditions. Crop Prot 64:110–114

Nurk L, Graß R, Pekrun C, Wachendorf M (2017) Effect of sowing method and weed control on the performance of maize (*Zea mays* L.) intercropped with climbing beans (*Phaseolus vulgaris* L.). Agriculture 7:51

Oerke EC (2006) Crop losses to pests. J Agric Sci 144:31–43

Oerke EC, Dehne HW (2004) Safeguarding production—losses in major crops and the role of crop protection. Crop Prot 23:275–285

Pannacci E, Tei F (2014) Effects of mechanical and chemical methods on weed control, weed seed rain and crop yield in maize, sunflower and soyabean. Crop Prot 64:51–59

Pannacci E, Lattanzi B, Tei F (2017) Non–chemical weed management strategies in minor crops: a review. Crop Prot 96:44–58

Panneton B, Guillaume S, Roger J-M, Samson G (2010) Improved discrimination between mono-cotyledonous and dicotyledonous plants for weed control based on the blue–green region of ultraviolet–induced fluorescence spectra. Appl Spectrosc 64:30–36

Pantazi X-E, Moshou D, Bravo C (2016) Active learning system for weed species recognition based on hyperspectral sensing. Biosyst Eng 146:193–202

Partel V, Kakarla SC, Ampatzidis Y (2019) Development and evaluation of a low-cost and smart technology for precision weed management utilizing artificial intelligence. Comput Electron Agric 157:339–350

Peñuelas J, Filella I (1998) Visible and near-infrared reflectance techniques for diagnosing plant physiological status. Trends Plant Sci 3:151–156

Peteinatos GG, Weis M, Andújar D, Rueda-Ayala V, Gerhards R (2014) Potential use of ground-based sensor technologies for weed detection. Pest Manag Sci 70:190–199

Peteinatos GG, Korsaeth A, Berge TW, Gerhards R (2016) Using optical sensors to identify water deprivation, nitrogen shortage, weed presence and fungal infection in wheat. Agriculture 6:24

Pot V, Benoit P, Le Menn M, Eklo OM, Sveistrup T, Kvaerner J (2011) Metribuzin transport in undisturbed soil cores under controlled water potential conditions: experiments and modeling to evaluate the risk of leaching in a sandy loam soil profile. Pest Manag Sci 67:397–407

Power EF, Kelly DL, Stout JC (2013) The impacts of traditional and novel herbicide application methods on target plants, non-target plants and production in intensive grasslands. Weed Res 53:131–139

Rasmussen J, Nielsen J, Streibig JC, Jensen JE, Pedersen KS, Olsen SI (2019) Pre–harvest weed mapping of *Cirsium arvense* in wheat and barley with off–the–shelf UAVs. Precis Agric 20: 983–999

Rosell JR, Sanz R (2012) A review of methods and applications of the geometric characterization of tree crops in agricultural activities. Comput Electron Agric 81:124–141

Roslim MHM, Juraimi AS, Che'Ya NN, Sulaiman N, MNHA M, Ramli Z, Motmainna M (2021) Using remote sensing and an unmanned aerial system for weed management in agricultural crops: a review. Agronomy 11:1809

Rueda-Ayala V, Rasmussen J, Gerhards R (2010) Mechanical weed control. In: Oerke EC, Gerhards R, Menz G, Sikora R (eds) Precision crop protection—the challenge and use of heterogeneity. Springer, Dordrecht, pp 279–294

Rueda-Ayala V, Peteinatos G, Gerhards R, Andújar D (2015) A non–chemical system for online weed control. Sensors 15:7691–7707

Rydahl P (2003) A web-based decision support system for integrated management of weeds in cereals and sugarbeet. Bull OPPE 33:455–460

Scotford I, Miller P (2005) Applications of spectral reflectance techniques in northern European cereal production: a review. Biosyst Eng 90:235–250

Shahbazi N, Ashworth MB, Callow JN, Mian A, Beckie HJ, Speidel S, Nicholls E, Flower KC (2021) Assessing the capability and potential of LiDAR for weed detection. Sensors 21:2328

Slaughter D, Giles D, Downey D (2008) Autonomous robotic weed control systems: a review. Comput Electron Agric 61:63–78

Søndershov M, Fritzsche R, de Mol F, Gerowitt B, Goltermann S, Kierzek R, Krawczyk R, Bøjer OM, Rydahl P (2015) DSSHerbicide: weed control in winter wheat with a decision support system in three South Baltic regions–field experimental results. Crop Prot 76:15–23

Sportelli M, Frasconi C, Fontanelli M, Pirchio M, Raffaelli M, Magni S, Caturegli L, Volterrani M, Mainardi M, Peruzzi A (2021) Autonomous mowing and complete floor cover for weed control in vineyards. Agronomy 11:538

Streibig JC, Rasmussen J, Andújar D, Andreasen C, Berge TW, Chachalis D, et al (2014) Sensor-based assessment of herbicide effects. Weed Res 54:223–233

Su WH (2020a) Advanced machine learning in point spectroscopy, RGB–and hyperspectral–imaging for automatic discriminations of crops and weeds: a review. Smart Cities 3:767–792

Su WH (2020b) Crop plant signaling for automated crop/weed identification: a systematic review and new concept. Artif Intell Agric 4:262–271

Su WH, Sun DW, He JG, Zhang LB (2017) Variation analysis in spectral indices of volatile chlorpyrifos and non–volatile imidacloprid in jujube (*Ziziphus jujuba* Mill.) using near–infrared hyperspectral imaging (NIR–HSI) and gas chromatograph–mass spectrometry (GC–MS). Comput Electron Agric 139:41–55

Sui R, Thomasson JA, Hanks J, Wooten J (2008) Ground–based sensing system for weed mapping in cotton. Comput Electron Agric 60:31–38

Swain KC, Zaman QU, Schumann AW, Percival DC (2009) Detecting weed and bare–spot in wild blueberry using ultrasonic sensor technology. American Society of Agricultural and Biological Engineers 096879, ASAE, St. Joseph

Tamouridou AA, Alexandridis TK, Pantazi XE, Lagopodi AL, Kashefi J, Moshou D (2017) Evaluation of UAV imagery for mapping Silybum marianum weed patches. Int J Remote Sens 38:2246–2259
Thorp KR, Tian LF (2004) A review on remote sensing of weeds in agriculture. Precis Agric 5:477–508
Tian L (2002) Development of a sensor-based precision herbicide application system. Comput Electron Agric 36:133–149
Torres-Sánchez J, Peña JM, de Castro A, López-Granados F (2014) Multi–temporal mapping of the vegetation fraction in early–season wheat fields using images from UAV. Comput Electron Agric 103:104–113
Travlos IS, Bilalis DJ, Katsenios N, De Prado R (2018) Sustainable weed control in vineyards. In: Korres NE, Burgos NR, Duke SO (eds) Weed control: sustainability, hazards, and risks in cropping systems worldwide. CRC Press, Boca Raton, pp 526–542
Travlos I, Tsekoura A, Antonopoulos N, Kanatas P, Gazoulis I (2021) Novel sensor-based method (quick test) for the in-season rapid evaluation of herbicide efficacy under real field conditions in durum wheat. Weed Sci 69:147–160
Tremblay N, Wang Z, Ma BL, Belec C, Vigneault P (2009) A comparison of crop data measured by two commercial sensors for variable-rate nitrogen application. Precis Agric 10:145–161
Tyystjärvi E, Nørremark M, Mattila H, Keränen M, Hakala-Yatkin M, Ottosen C-O, Rosenqvist E (2011) Automatic identification of crop and weed species with chlorophyll fluorescence induction curves. Precis Agric 12:546–563
Upadhyaya SK, Pérez-Ruiz M (2012) GNSS in precision agricultural operations. Intech Open, London
van Evert FK, Samsom J, Polder G, Vijn M, Dooren HJV, Lamaker A et al (2011) A robot to detect and control broad-leaved dock (Rumex obtusifolius L.) in grassland. J Field Robot 28:264–277
Van Evert FK, Gaitán-Cremaschi D, Fountas S, Kempenaar C (2017) Can precision agriculture increase the profitability and sustainability of the production of potatoes and olives? Sustain For 9:1863
Wang A, Zhang W, Wei X (2019) A review on weed detection using ground–based machine vision and image processing techniques. Comput Electron Agric 158:226–240
Wei X, Bai J, Zhang J, Wu S (2014) Weed identification from winter rape at seedling stage based on spectral variable selection. Sens Lett 12:530–536
Wiltshire JJJ, Tillett ND, Hague T (2003) Agronomic evaluation of precise mechanical hoeing and chemical weed control in sugar beet. Weed Res 43:236–244
Xu Y, Gao Z, Khot L, Meng X, Zhang Q (2018) A real-time weed mapping and precision herbicide spraying system for row crops. Sensors 18:4245
Yang C-C, Prasher SO, Landry J-A, Ramaswamy HS (2003) Precis Agric 4:5–18
Zhang Y, Slaughter D (2011) Hyperspectral species mapping for automatic weed control in tomato under thermal environmental stress. Comput Electron Agric 77:95–104
Zhang CJ, Lim SH, Kim JW, Nah G, Fischer A, Kim DS (2016) Leaf chlorophyll fluorescence discriminates herbicide resistance in *Echinochloa* species. Weed Res 56:424–433
Zhang Y, Gao J, Cen H, Lu Y, Yu X, He Y, Pieters JG (2019) Automated spectral feature extraction from hyperspectral images to differentiate weedy rice and barnyard grass from a rice crop. Comput Electron Agric 159:42–49
Zheng Y, Zhu Q, Huang M, Guo Y, Qin J (2017) Maize and weed classification using color indices with support vector data description in outdoor fields. Comput Electron Agric 141:215–222
Zimdahl RL (2018) Fundamentals of weed science, 5th edn. Academic, Cambridge, MA

Chapter 6
Precision Irrigation in Outdoor Horticulture

Ahmed G. El-Naggar, Carolyn B. Hedley, and Pierre Roudier

Abstract Effective management of cropping and horticultural systems depends on effective management of water resources. Water availability and impacts on water quality are increasingly problematic for irrigated arable cropping and outdoor horticulture and require more efficient use of water resources. Growers need to demonstrate efficient and sustainable use of water to remain competitive.

Developing innovative site-specific management approaches, such as precision irrigation, that combine knowledge of the whole system—soil, crop and irrigation performance—with spatio-temporal water applications can improve crop water productivity and energy-savings with less impact on the environment, when well designed and well managed. The aim of this chapter is to discuss the latest state-of-the-art technologies that enable precision irrigation management systems, with examples of their use for outdoor horticulture. The chapter includes:

- a definition and an overview of precision irrigation,
- a review of sensing technologies and their integration into precision irrigation decision support systems, including variable rate delivery technologies that optimise water, energy consumption and other environmental impacts, and
- identification of current research in precision irrigation application for horticulture crops, including barriers and future research in precision irrigation.

A. G. El-Naggar (✉)
Land and Water Management Department, IHE Delft Institute for Water Education, Delft, The Netherlands
e-mail: a.elnaggar@un-ihe.org

C. B. Hedley
Manaaki Whenua – Landcare Research, Massey University Campus, Palmerston North, New Zealand

P. Roudier
Manaaki Whenua – Landcare Research, Massey University Campus, Palmerston North, New Zealand

Te Pūnaha Matatini, A New Zealand Centre of Research Excellence, Auckland, New Zealand

P. Newell Price et al. (eds.), *Precision Agriculture in Field Horticulture*, Progress in Precision Agriculture, https://doi.org/10.1007/978-3-031-96534-0_6

Keywords Precision irrigation · Precision horticulture · Variable rate irrigation · Water use efficiency · Soil sensing

6.1 Introduction

Continued population growth and depletion of freshwater resources are creating significant pressures on agriculture around the world. This is exacerbated by climate change that is expected to bring more extreme climatic conditions including flooding and droughts. Projected increases in water use will increase the cost and decrease the availability of freshwater resources for agriculture.

Irrigated production systems produce nearly 40% of the world's food and consume about 80% of total freshwater withdrawals (FAOSTAT data 2022). Thus, as the population grows, new and more sustainable solutions for the production of food will be of key importance for irrigated horticulture and agriculture. This will require systems and methods to monitor, measure and respond to growing conditions, forecast water demand and strategically schedule water allocation based on available water supplies. These requirements are global, and even countries that have relatively abundant freshwater, such as New Zealand and Canada, need to address regional or temporal water shortages due to uneven rainfall distribution (El-Naggar 2020).

Pressurized irrigation systems, such as continuous move irrigation systems (primarily centre-pivots or linear moves) and fixed irrigation systems (solid set sprinklers and localized drip irrigation), deliver water in smaller, more precise amounts than surface flood irrigation systems that leads to opportunities for improved water use efficiency (WUE) when well managed. When pressurized systems have variable rate control then factors, such as soil spatial variability, can be taken into consideration and water placement can be further fine-tuned.

Over-irrigation contributes to increased drainage and surface runoff with the accompanying risk of nutrient leaching to groundwater, which wastes irrigation water and fertiliser and decreases water quality. It also increases risk of waterlogging with the accompanying increased risks of nitrous oxide emissions, reduced yield and quality, increased cost of irrigation and reduced profit. Under-irrigation causes reductions in yield and quality and lower overall WUE. It can also lead to salinity issues in arid climates when salts (or dissolved minerals) in the irrigation water are not sufficiently flushed through the soil surface. This is because under-irrigation allows these salts, which are naturally present in most water sources, to accumulate in the root zone instead of being flushed out.

Increasing global demand for high-quality horticultural products, coupled with the increasing costs of crop production and farm operation, requires irrigation systems to shift from conventional systems to those with more precise control (e.g. variable rate and placement), referred to here as 'precision irrigation' (PI). The development of site-specific water management tools, such as PI equipment, has great potential to increase crop water productivity and reduce nutrient leaching losses (over-irrigation) or build-up of salinity (under-irrigation). Farmers and growers need to make decisions on how to optimize water use with crop and fruit production, to remain competitive. This chapter will discuss the development of PI and its associated sensing technologies and decision support system (DSS) technologies. It also provides some examples of the actual use of PI in outdoor horticulture cropping situations.

6.2 Precision Irrigation

Overview of Precision Irrigation

Precision irrigation is a method of adapting and applying irrigation water to fit the site- and time-specific water requirements of plants. The underlying rationale for PI is that there are often fine-scale spatial variations in plant irrigation needs across a production area (field, orchard or vineyard) because of (1) the spatial variability in soil properties (e.g. texture, available water-holding capacity and infiltration and drainage rate), (2) variable topography impacting water redistribution in the landscape, (3) spatial variations in crop demands and/or (4) variable irrigation performance. The main goal of a PI system is to adapt irrigation recommendations at finer spatial and temporal scales by continuous monitoring, thus optimising economic and environmental outcomes that include maximising WUE, optimising yield and reducing negative environmental impacts.

An example of a system that can currently achieve these goals is a variable-rate irrigation (VRI) system. Fitted with Global Navigation Satellite System (GNSS) technology, a VRI system can apply variable irrigation amounts, according to continuous monitoring of soil and crop condition and irrigation performance.

The PI paradigm is technically applicable to all irrigation application methods and for all crops, although different methods will have varying constraints related to the potential spatial and temporal scales of application. Research into PI has evolved since early studies were first reported in the USA in the early 1990s e.g. Fraisse et al. (1993). While a significant portion of recent work focuses on Variable Rate Irrigation (VRI) with continuous-move systems (e.g., El-Naggar et al. 2020; Hedley and Yule 2009b; King et al. 1999; Sadler et al. 1996, 2000; Wall et al. 1996), research has also been conducted on other irrigation methods. Studies have also been made of drip-irrigation systems, primarily in horticultural crops (e.g. Capraro et al. 2008; Ooi et al. 2008) and tree fruits (e.g. Coates and Delwiche 2008; Ortuani et al. 2019; Sanchez et al. 2017). Automation and control of surface flood systems has been advanced and shown to provide a higher level of irrigation performance along with substantial labour savings than conventional surface systems (Hedley et al. 2014; Khatri and Smith 2007). Other systems, such as solid set sprinkler and travelling gun systems, have received the least attention to date in PI, most likely because of the lower precision in irrigation uniformity and application depths and the popularity of centre pivots and drip systems.

Benefits of Precision Irrigation

There have been numerous comprehensive *in silico* (simulation) studies evaluating the benefits of PI for agricultural production and environmental outcomes in various sectors including horticulture (see Sect. 6.5) e.g. (Daccache et al. 2015; DeJonge et al. 2007; El Chami et al. 2019; Hedley and Yule 2009b; Nijbroek et al. 2003; Perea et al. 2018) along with *in vivo* (field trial) experiments e.g. (El-Naggar et al. 2020; Irmak and Rathje. 2014; Lo et al. 2014; Sadler et al. 2005). The potential benefits of PI include:

- **Reduced environmental footprint**. Lo et al. (2014), for example, suggested that the splitting of irrigation applications or reduction of their intensity using PI methods may enable a decrease in the gross irrigation and runoff with improved infiltration into areas of high potential runoff risk.
- **Optimised water use efficiency**. Evapotranspiration (ET_c) may be spatially heterogeneous when crops of diverse planting dates, maturity lengths, and/or species are grown concurrently on one field and PI can be adjusted to match these differences in crop water use. For example, Sadler et al. (2005) concluded that variable rate irrigation (VRI) could save 10–15% of water used in conventional irrigation practice. Hedley and Yule (2009b) also reported water savings of around 25% with VRI adjusted to suit the available water-holding capacity (AWC) of the soils. Areas with no crop can remain unirrigated altogether.
- **Manage fertiliser application**. Sadler et al. (2005) suggested that precision chemigation and effluent application can be practiced on fields where the law prohibits agrochemical and effluent losses to the water bodies within the field. Also, some agrochemical applications may be more cost effective with spatio-temporal chemigation than with other application methods.
- **Improved yield**. Irmak and Rathje (2014), for example, found PI was a desirable method to avoid over-irrigation, which can reduce yield due to the promotion of plant diseases, the decrease in nutrient availability, and limited root growth and function.

Components of a Precision Irrigation System

There are four essential, interlinked steps in the process of PI. Appropriate integrated technologies are required for: monitoring, recommendation, application, and evaluation.

- **Monitoring**. A PI approach includes evaluating any spatio-temporal variability in soil and crop conditions and irrigator performance within and between the production area e.g. field, orchard or vineyard. It does this by identifying and quantifying such variability using high resolution sensor monitoring and data acquisition systems (See Sects. 6.3.1 and 6.3.2).

- **Recommendation**. The collected data at the monitoring step needs to be interpreted and analysed at an appropriate scale. The latter has been identified as a major constraint to the adoption of PI so developing multi-dimensional DSS that fully automate real-time irrigation in a cost-effective manner is essential (See Sect. 6.4).
- **Application**. The irrigation recommendations produced based on monitoring data are then applied by the irrigation system. Variable-rate delivery systems are designed to optimise water use in two ways: (1) by varying the application rate and/or (2) by varying the application time. Automatic controllers with real-time data from on-the-go sensors should provide the most reliable and potentially accurate means of controlling irrigation applications (See Sects. 6.3.2 and 6.4).
- **Evaluation**. Measurement of the engineering, agronomic and economic performance of the irrigation system is essential for providing feedback and improvement for the next irrigation cycle (See examples in Sect. 6.5).

6.3 Sensing Technologies Associated with Precision Irrigation

As mentioned, the underlying rationale for implementation of PI technologies is the need to manage any spatiotemporal variability within a production area that affects the irrigation needs of a crop. Any variability in crop water needs within the field, orchard or vineyard are due to: (1) soil and land variability such as different soil depths, textures, AWC, infiltration rates, drainage rates and microtopography effects, (2) staged planting dates and variable plant growth, (3) irrigator performance issues, e.g. blocked sprinklers, leakage (e.g. in drip irrigation systems).

The data collected by (i) on-the-go or in-field fixed ground-based (proximal) sensors, (ii) sensors on airborne platforms (e.g. unmanned aerial vehicles, small fixed wing aeroplanes) and (iii) spaceborne sensors (installed on satellite platforms, such as Landsat, Spot, Sentinel, MODIS, etc.) show huge promise for monitoring this spatial variability of soil, crop and irrigation performance. The sensing information is useful to guide soil and crop sampling and to assist decision-making processes regarding PI management. An effective method for converting large volumes of collected sensor data to useful information is the key to the success of these systems.

The sources of sensing data commonly used in precision horticulture are categorised into proximal sensing, airborne sensors, and spaceborne sensors. In this chapter, we use the term "PS" for proximal (ground-based) sensors and "RS" for remote (e.g. airborne and satellite) sensors.

The majority of PS and RS tools measure the electromagnetic reflectance from a surface across several wavelengths (Fig. 6.1). In comparison with ground-based sensors, airborne and spaceborne sensors have poorer resolution but larger ground coverage. Spaceborne sensors, however, have seen the spatial resolution of their

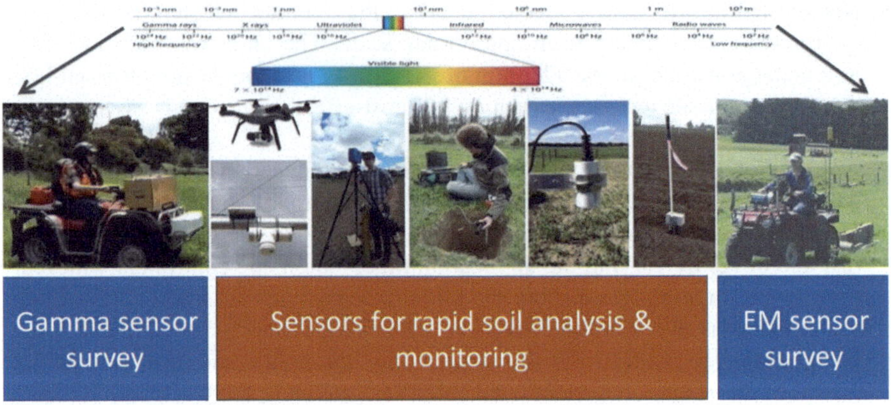

Fig. 6.1 The Electromagnetic Spectrum: Sensor systems designed to capture specific ranges of the electromagnetic spectrum

imagery drastically improved throughout the last two decades, with the Sentinel satellite constellation now offering 10 m resolution imagery and Planet providing 5 m resolution imagery. Additionally, another revolution is the free availability of high-resolution earth monitoring missions (both optical and radar), such as Landsat and Sentinel. As the affordability and accessibility of PS and RS data sources and information provision improves, these technologies are increasingly becoming more available for horticulture crops, orchards and vineyards for mapping, tree monitoring, harvest optimization and input management.

Monitoring Spatial Variations for Soil and Crop Characteristics

Mapping Soil Spatial variability Using Proximal Sensors

Site-specific water management is strongly influenced by soil properties, especially soil texture. Therefore, a sensor map that relates to soil texture or soil water content (SWC) with sufficient resolution is very useful to effectively partition PI management zones.

New geophysical technologies for mapping soil properties, specifically ground-penetrating radar (GPR), electrical resistivity (ER), or electromagnetic induction (EMI) can be used to effectively interpret soil variability within the production area that can guide PI (e.g. Brevik et al. 2006; Corwin and Lesch 2005; El-Naggar et al. 2020; Hedley and Yule 2009a; Hedley et al. 2004; Huang et al. 2017). In comparison with traditional physical soil sampling methods, geophysical technologies can produce high-resolution data and reduce costs, increasing time and cost efficiencies.

Ground-penetrating radar (*GPR*) is a non-invasive method that sends electromagnetic energy into the ground and measures the time required for the wave to return to a receiving antenna. The amount of energy that is reflected back to the receiver, the strength of the reflection, is determined by the difference in dielectric constant between the object or layer of interest and the surrounding medium. GPR has been extensively used for various applications, including PI, for example, (1) the determination of SWC (e.g. Grote et al. 2010; Huisman et al. 2003) and (2) subsurface free water and hydraulic properties (e.g. Lubczynski and Roy 2004; Lane 2016). Furthermore, Hubbard et al. (2002) used high-frequency GPR ground wave data to map spatial and temporal variations in SWC of a vineyard. Duncan and Leib (2011) used GPR to identify spatial variability of AWC for PI management of a deficit irrigation study. However, GPR is an imaging tool that needs expert interpretation skills.

The following sensor survey methods are used more commonly for precision horticulture applications. They provide numeric datasets which can be analysed to delineate variability in the field.

Gamma Radiometrics is designed to detect the gamma rays naturally emitted from the Earth's surface and is used for soil variability mapping (e.g. Wong and Harper 1999). It can be considered as a direct, integrated measure of the mineralogical and textural composition of the soil, and it has been applied to estimate variations in surface SWC (e.g. Rossel et al. 2010). However, the relationship of gamma survey data to soil moisture is not as direct as that of electrical resistivity (ER) and electromagnetic induction (EMI) techniques, described below.

ER represents the capacity of soil materials to resist the flow of electrical current. Samouëlian et al. (2005) described the apparent soil ER as a complex function of the composition and arrangement of solid soil constituents, porosity, pore-water saturation, conductivity of the pore-water, and temperature. ER methods can be divided into those that inject currents into the ground through direct coupling and those that do this through capacitively induced coupling. Typically, both methods measure the apparent ER, which is subsequently converted to its inverse, the apparent electrical conductivity (EC_a) of the soil. A range of PI studies has investigated the use of ER to map spatial water content within irrigated fields (e.g. Samouëlian et al. 2005). Recently, Mertzanides et al. (2020) found spatio-temporal soil ER data revealed a strong relationship with gravimetric soil moisture point measurements. It is an invasive technique in contrast to EMI which can be used for rapid non-invasive soil sensor survey mapping.

EMI is a highly adaptable non-invasive technique that measures the apparent bulk EC_a of the soil. The principle of EMI sensors is to firstly generate a primary magnetic field that induces very small currents in the soil that in turn generate a secondary magnetic field. This secondary magnetic field is measured by a receiver coil in the sensor. Sensors are designed so that the secondary and primary magnetic field is linearly proportional to soil EC_a. Measurement of soil EC_a is a frequently used and a very reliable method to indicate spatial soil variability for precision agriculture (Corwin and Lesch 2003). The depth of exploration of the soil profile is proportional (for homogeneous material) to the distance between the transmitting

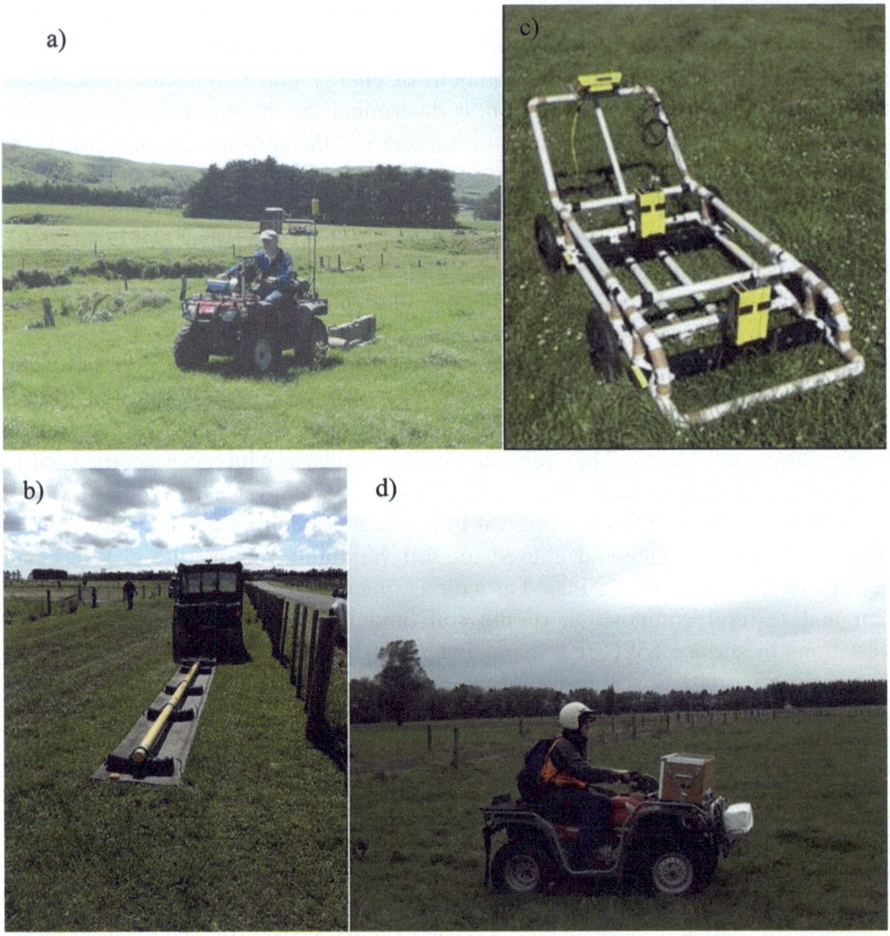

Fig. 6.2 Four geophysical technology sensors commonly used in soil investigations are: Electromagnetic induction (EMI) (**a**) EM38, (**b**) DUALEM421, (**c**) Ground-penetrating radar (GPR) and (**d**) Gamma radiometrics

and sensing coils for EMI sensors. Examples of EMI sensors are shown in Fig. 6.2. Multi-coil EMI sensors, e.g. DUALEM-421 (Dualem Inc., Milton, Ontario, Canada), incorporate an EM transmitter that operates at a fixed low frequency (9 kHz) and 3 pairs of horizontal co-planar (HCP) and perpendicular (PRP) receiver arrays. The depth of EC_a measurement is, respectively, 0–1.5 (1mHcon), 0–3.0 (2mHcon) and 0–6.0 m (4mHcon), and 1.1, 2.1 and 4.1 m (PRP) (El-Naggar et al. 2021a). The EC_a value is dependent on the electrical conductivity of the soil which is largely controlled by moisture and clay content in non-saline soils. Studies also exist that relate EC_a to cation exchange capacity, bulk density, the presence of pans, clay mineralogy and soil temperature (e.g. Doolittle et al. 1994; Sudduth et al. 2001).

There are many examples of studies that include the use of EC_a survey data to delineate spatial soil variability (PI zones) in addition to soil properties. Sudduth et al. (2005) concluded that soil EC_a can be calibrated to any specific soil property so long as the soil property of interest dominates the other local soil properties that influence EC_a readings. Doerge et al. (1999) suggested that EC_a could be successfully related to AWC, which would be closely related to soil texture, so long as proper field verification is performed. EC_a is also known to relate well to clay content (e.g. Sudduth et al. 2005; Williams and Hoey 1987). Sand particles have a low EC_a while the charged surfaces of clay particles have a high conductivity; therefore, finely textured soils should produce high EC_a values, while coarse soils should produce low EC_a values.

Hezarjaribi and Sourell (2007) used EC_a to predict AWC. This technique has been implemented using a Geonics- EMI instrument (Geonics Ltd. https://www.geonics.com/) or a Veris coulter ER type sensor (Veris Technologies Inc. https://www.veristech.com/). Hedley and Yule (2009a) also used an indirect approach to first delineate the field into management zones based on the dense EC_a survey data and then assigned a uniform AWC to each management zone based on the AWC of the sampled locations within each management zone. Regardless of the approach, a strong relationship between AWC and EC_a is critical to the success of this technique for making AWC maps. If such a relationship does not exist in the field of interest, then other dense geospatial datasets, including historical yield maps, would be needed. The success of the research mentioned above that relates EC_a to AWC and soil texture characteristics indicates that EC_a could potentially allow users to successfully identify precision management zones for PI management.

Mapping Crop Parameters Using Optical and Multi-spectral Sensors

By measuring the appropriate plant parameters, it is possible to evaluate a plant's general health and to use this to inform the application of water and fertilisers in the right place at the right time. A wide range of multi-spectral and optical sensors have demonstrated capabilities for making non-invasive measurements of crop canopies to characterize crop growth and identify related key variables at finer and finer spatial resolutions as sensor technologies improve. A wide range of PS and RS technologies now exist to monitor plant growth, but here the focus is on three that offer potential for greater operational inclusion into DSS for PI. These are: (1) crop sensors that measure Normalised Difference Vegetation Index (NDVI), (2) infrared radiometers and (3) light detection and ranging sensors.

Crop Sensors That Measure Normalised Difference Vegetation Index

Many vegetation indices have been created using spectral data from a wide range of crop sensors to detect the health of vegetation. Vegetation indices are effective in monitoring crop growth conditions from emergence to full cover and are influenced by variable field or orchard conditions e.g. tillage, irrigation, fertilizer application etc. Vegetation indices are used to monitor excessive moisture in fields, to detect drought areas and assess the weather impact on vegetation growth and crop production (e.g. Kogan et al. 2003; Singh et al. 2003).

The normalised differences vegetation index (NDVI) has gained wide acceptance primarily due to its ease of use, its use of only two wavelengths, and the important plant characteristics the ratio reflects. NDVI is comprised of a ratio of the difference between reflectance of near-infrared and red radiation. The underlying notion is that red radiation (650 nm) is absorbed and near-infrared (NIR) radiation (700–1300 nm) is highly reflected by healthy tissue.

NDVI is a potentially useful indicator of water stress and canopy cover because irrigated crops should stay greener longer than non-irrigated crops. Abuzar et al. (2001) found that NDVI and plant moisture availability are closely related and when plant moisture is the major varying factor then NDVI is a sufficiently good indicator for irrigation need According to Plant and Keeley (1999), there are many reasons why NDVI may relate to water stress: soil water ultimately has a cumulative effect on many vegetation properties, such as the quality of the canopy structure as impacted by wilting or possible reductions in leaf quantities.

NDVI has proven to be extremely useful for mapping irrigated areas versus non-irrigated areas mainly due to the differential spectral responses between irrigated and non-irrigated croplands. Research studies, e.g. Ozdogan et al. (2010) and Plant and Keeley (1999), have shown that plant growth and development can be mapped effectively with remotely sensed reflectance data and numerous studies have shown high correlations between vegetation indices developed from spectral observations and plant stand parameters (Wiegand et al. 1991). Studies by Biggs et al. (2006) and Ozdogan et al. (2006) have shown the ability to account accurately for irrigated land-use in other semi-arid to arid regions of the world using NDVI.

Several studies have tested the potential use of NDVI to predict site-specific crop coefficients and ET_c for a broad range of crops at large and small scales, for example D'Urso et al. (2008), El-Naggar et al. (2020), El-Naggar et al. (2021b) and González-Dugo and Mateos (2008). This approach has the potential to capture finer resolution spatial variability of water use in irrigated crops and for wider adoption by PI DSSs.

Methods for Measuring Plant Water Use Including Infrared Thermometry (IRT)

It is believed that neither soil water status nor the atmospheric demand accurately represents the plant water status as well as the plant itself. However, direct measurement of daily water use by the plant is not easy (as discussed in this section).

According to White and Raine (2008), plant water status can be determined by measuring the plant's response to a change in tissue water status. Plant-based sensing may measure a direct physiological indicator (e.g. plant water status) or an indirect physiological plant response induced by changes in plant water status (e.g. leaf temperature, plant organ diameter or growth). The most frequently used methods for automatic irrigation control are the canopy temperature (Tc) method and the sap flow method.

A limitation of plant-based methods is that they tend to indicate the day on which the plant is going into stress and needs irrigation, i.e. they are not predictive. In comparison, with a knowledge of soil AWC, soil moisture measurements can be used to determine how many days of available water is stored in the soil and to predict the number of days before irrigation is required. This information, together with climate data, is critically useful for PI DSS, and especially relevant in temperate climates where some rainfall is expected during the irrigation season.

Canopy Temperature (Tc) Method

The use of Tc readings has been shown to reflect subtle changes in physiological processes, such as cell growth and biochemical reactions associated with the damaging effects of super optimal temperature (Conaty 2010). Jones (2004) confirmed that irrigation scheduling can be improved by monitoring crop Tc using IRT. When IRT sensors are mounted in the field then Tc can be used to directly measure and map the temperature variability of a crop. The spatial variability in Tc should be low for a well-watered, non-stressed crop but increase as plant water stress increases, for variable crop, soil and irrigator conditions. By capturing real-time variations in Tc across the field, IRT sensors provide valuable data for PI systems to optimize irrigation decisions and target water application to specific crop needs. This technology has been implemented in various studies, for example Peters and Evett (2008) installed IRT sensors on irrigation machines for on-the-go data capture.

In addition, the difference between Tc and air temperature (Tc-Ta) is related to plant water stress (Widmoser 2010). The Tc-Ta index was first studied by Ehrler (1973), who investigated the possibility of using Tc-Ta as a guide for irrigation scheduling. Furthermore, Idso et al. (1981) observed a linear relationship between Tc (measured using IRT) and Ta and vapour pressure deficit, which they used to develop an empirical method for quantifying crop water stress. Jackson et al. (1981) also conducted theoretical research to develop a crop water stress index (CWSI). These two approaches are based on the principle that transpiration cools the leaf surface and as water becomes limited, stomatal conductance and transpiration decrease, leading to increases in leaf temperature. However, given that ambient conditions can have a large influence on Tc, then Tc values are in fact a reflection of both plant and environmental factors (Conaty 2010; Jones 2006). Both approaches have been proposed to estimate the lower limiting Tc by using two baselines (non-water stressed, and water-stressed). The lower baseline represents Tc–Ta (denoted as Tc-Ta) of a well-watered crop transpiring at the maximum potential rate while the upper baseline represents Tc-Ta of a non-transpiring crop (Tc-Ta_non-transpiring).

Worldwide, CWSI has been used as a tool to indicate plant water status and for scheduling irrigation in many crops, for example Cremona et al. (2004), Erdem et al. (2010) and Yildirim et al. (2012). Nonetheless, for CWSI to be an effective tool for scheduling irrigation, it has to be determined for particular crops in specific climates, given that crop response to water stress depends on local environmental conditions (Yildirim et al. 2012). Erdem et al. (2010) and Köksal (2008) have evaluated the application of CWSI in irrigation scheduling for different horticulture crops, including vegetables. Canopy to air temperature differences have been correlated to SWC and stem water potential for potential use in managing site-specific irrigation of peach orchards (Wang and Gartung 2010).

Infrared-thermometry and thermography can be used to indirectly estimate stomatal aperture as it responds to water stress. Measuring changes in stomatal conductance is an alternate, but more difficult, method to indicate crop water stress. Photosynthetic uptake of carbon dioxide and transpiration are both controlled by the opening and closing of stomata. Jones (2006) stated that measurements of stomatal conductance are sensitive to declining soil water and thus, water stress. These approaches have the advantage of viewing a larger area than crop models as well as being capable of automation for real-time measurements. However, calculating stomatal conductance from Tc is difficult and sensitive to other environmental factors, such as radiation, Ta and wind velocity as well as leaf size (Scherrer et al. 2011), and the method needs further development before it could be utilised operationally by farmers and growers.

Sap Flow Method

Sap flow measurements also provide a unique and valuable way to quantify the time course of whole-plant water use under production conditions. The method applies heat to the stem and detection of the rate of sap flow or sap-flux density by monitoring the stem's thermal regime. A sap flow research method has been developed to measure water use across a wide range of applications on diverse plant types, including natural and urban forest trees (Čermák et al. 2004), woody horticultural trees and vines (e.g., Alarcón et al. 2003; Fernández Luque et al. 2008; Green and Clothier 1988), and agricultural crop species (Cohen et al. 1988). The 'compensation heat-pulse method' of Green and Clothier (1988) has been shown to be a useful indicator of transpiration from young apricot trees and, although the two factors (sap flow and transpiration) deviate at times, the method does enable a useful approximate automated recording method for transpiration (Alarcon et al. 2003). Commercial sap flow sensors are available and there is potential for spatial sap flow monitoring for PI.

Light Detection and Ranging (LiDAR)

LiDAR is a non-contact optical device that accurately measures the distance to a target based on the elapsed time between the emission and return of laser pulses (the time-of-flight method) or based on trigonometry (the optical-probe or light-section methods) so that precision 3-D information about the target can be obtained (Hosoi and Omasa 2009; El-Naggar et al. 2021b). The time-of-flight method is suitable for both short and far range measurements, while triangulation sensors measure short ranges (maximum a few meters) with high accuracy (Ehlert et al. 2009b).

LiDAR systems can be incorporated into (i) satellites and airborne systems (ALS) for detecting larger areas of interest e.g. whole farms from fixed-wing aircraft (200–1000 m altitude; Dalla Corte et al. (2020)), or (iii) terrestrial or ground-based LiDAR sensors (TLS) which are suitable for surveying purposes for forest ecology and precision agriculture (spatial resolution of 10–50 mm).

Advantages of the LiDAR optical-sensing method include its accuracy, and densely sampled elevation point datasets. A main limitation for LiDAR-based technologies for land surface surveys is vegetation cover density, where too small gap fractions in the canopy prevent the laser pulse from reaching the ground. Its potential use for crop height mapping is still to be realised in PI systems.

Comprehensive precision agriculture-related studies confirm LiDAR is technologically and economically workable. For example, Hofle (2014) used LiDAR for individual maize plant detection; Saeys et al. (2009) used LiDAR to estimate the density of small grains; and Zhang and Grift (2012) measured crop heights for different crops. Ehlert et al. (2009a, 2010c, d) conducted various studies for measuring crop biomass with the use of different LiDAR sensors.

Little work has been undertaken to examine the use of LiDAR in PI applications. Tsoulias et al. (2019) used a LiDAR system for investigating the impact of the leaf area and soil patterns on water needs in an apple orchard. The study concluded that LiDAR-based plant data are potentially useful for optimizing and predicting monthly water needs spatially within orchards. Research presented by El-Naggar et al. (2021b) developed a method of using terrestrial LiDAR-based plant data combined with NDVI data to improve spatio-temporal estimates of the crop coefficient (Kc) and ETc. The method showed significant potential for improved high resolution (in time and space) monitoring of crop growth (height and biomass) to critically improve model-based prescription maps that underpin crop management decisions, such as VRI scheduling decisions.

Sensor Data Acquisition and Analysis Systems

PI management requires real-time monitoring indicators that determine when, where and how much to irrigate. The use of soil and crop sensors (discussed in Sect. 6.3.1) with wireless sensor networks (WSNs) installed into the field, orchard or vineyard, can provide real-time feedback on plant and soil water conditions by capturing the full range of variations (differences in moisture levels, temperature, etc.) across the entire area.

A number of WSNs with various topologies (e.g. star, mesh and long range (LoRa) networks) have been developed and investigated by researchers over the past decade. This includes WSN's for irrigation scheduling in centre-pivot irrigation, for example Barker et al. (2018); Ekanayake and Hedley (2018); Hedley et al. (2012); O'Shaughnessy and Evett (2008) and linear-move irrigation systems (Kim et al.

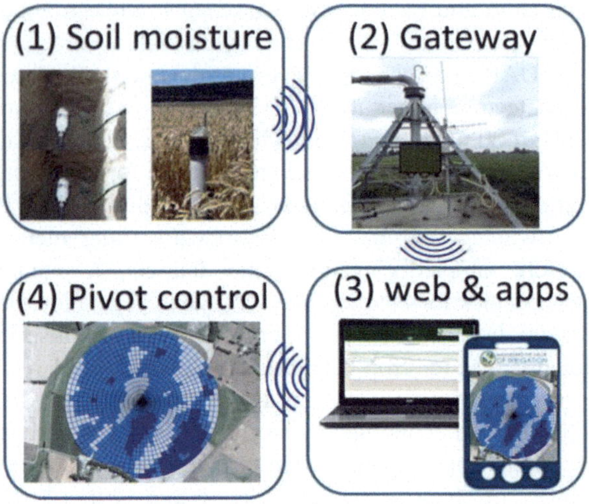

Fig. 6.3 Precision irrigation cycle of sensor data acquisition, analysis, and control system

2008). The recent rapid advance in wireless sensor networking technology has led to the development of readily accessible, low power, affordable, multifunctional sensor networks. The architecture of WSNs consists of a sensor attached to a node placed in a field that wirelessly sends the sensed data, for example soil moisture, to a base station. Then using cellular or internet connections the base station sends data to a cloud database where it is stored, processed and made available for a decision irrigation management tool, such as control software or a phone app (see an example in Fig. 6.3). Each sensor node comprises a sensor/actuator, controller, communication device, memory, and power supply.

An example of using WSN technology for scheduling in a VRI system is a study conducted by El-Naggar et al. (2020) where a WSN-soil moisture sensor-based method (see Fig. 6.4) was used to implement a VRI scheduling plan via a web-based user interface. The scheduling was conducted on pea and French bean crops grown under one centre pivot, with two delineated irrigation management zones. The sensor-based technique irrigated 23–45% less water compared to a model-based approach.

Other examples of the use of WSNs to transport data from sensors in the field to a remote database, where the data are analysed and converted to useful information for irrigation scheduling, are provided by Hedley et al. (2012) and O'Shaughnessy and Evett (2010). Hedley et al. (2012) developed a WSN soil moisture monitoring method under a centre pivot irrigator with VRI modification. This WSN method provided a direct measure of the SWC status every 15 min through a 3G cellular modem via the internet to a webpage. O'Shaughnessy and Evett (2010) developed a system that varied irrigation rate based on the crop factor, T_c. The system used a wireless IRT mounted on centre pivot irrigation arms as well as in the field to help determine VRI management.

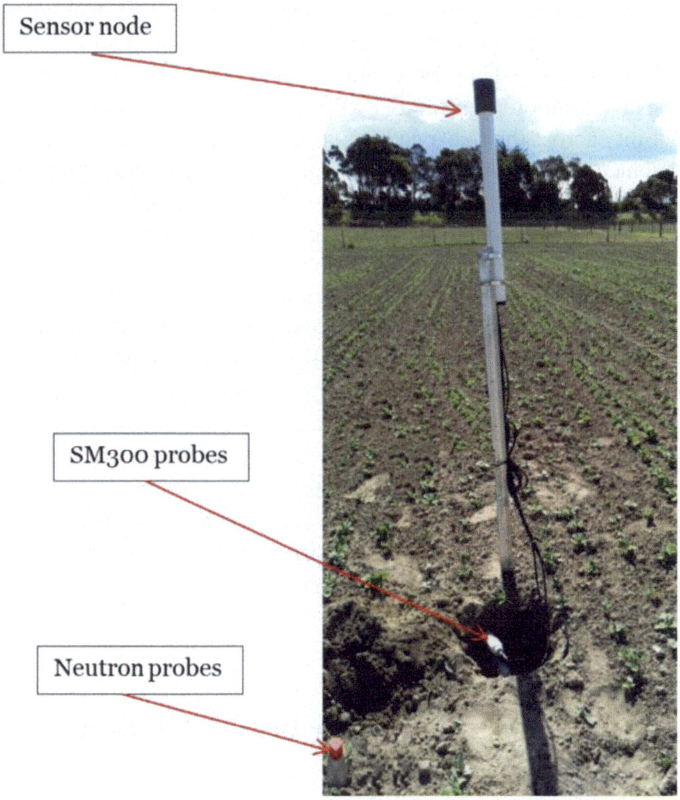

Fig. 6.4 Wireless soil moisture sensing system at Massey University No.1 Farm. The system provides soil moisture status in real-time. Soil moisture measurements from a neutron probe access tube was used for calibrating SM300 sensors. (El-Naggar et al. 2020)

Das et al. (2010) developed a WSN and a web-based application for in-field monitoring of a grape crop where data collected by the sensors were soil pH, SWC and soil temperature. If SWC dropped below a pre-determined threshold level, water application started. This threshold level must be decided based on climate, topography, and crop type. Similarly, Ooi et al. (2008) developed and tested automated WSN irrigation control systems for micro-irrigation in a large commercial vineyard. The scheduling was soil-moisture and ET based.

In another study, Anguraj et al. (2021) proposed a real-time WSN- soil moisture data collection system where irrigation is controlled according to collected data. The data were rapidly processed using a data mining method, a Hybrid neural network classifier, to produce efficient PI management. In addition, Jiménez-Buendía et al. (2021) illustrated in a recent study the feasibility of deploying a high-density network of nodes with a variety of heterogeneous sensors to collect data from the soil, plant, and atmosphere for a deficit PI application.

In summary, a large body of research shows that sensing technology has the ability to quantify water stress and aid PI scheduling for outdoor horticulture crops (annual and perennial) where PI management zones can be identified through proximal sensing technology such as EMI technology. Then temporal variations in water availability across the field can be monitored with soil moisture measurements, soil water balance (SWB) calculations, or plant water stress indicators. Soil moisture sensors and SWB remain the basis for PI management for annual and perennial cropping. However, there are virtually no Kc values in SWB models for ornamental species and most estimates of perennial crop water use are quite variable. Therefore, a SWB approach works best where it is combined with regular monitoring techniques, for example of SWC and CWSI, that can help reset the model.

The research indicated that a canopy temperature-based index (e.g. CWSI) is a sensitive indicator of water stress for perennial crops and, therefore, the WUE of these crops can be improved if irrigation is scheduled when water stress reaches a certain threshold. Recently, crop sensing technology methods, such as LiDAR combined with NDVI, have been used with annual horticulture crops to provide precise spatio-temporal information on the crop height and ground cover which can be used to recalibrate crop model estimates for non-optimal conditions to inform PI to achieve a better WUE, leading to greater yields, less water use, and lower nutrient losses.

6.4 Decision Support Systems for Precision Irrigation

To increase the adoption and feasibility of PI in outdoor horticulture, it is necessary to develop smart high-tech DSS that fully automate real-time, variable-rate application in a cost-effective manner. Coupling control software with a database that links to real-time soil and crop sensor data (Sect. 6.3.2) will provide the best irrigation management practises and will be more realistic and acceptable to both farmers and regulators.

Some studies have been conducted and research continues to develop DSSs for PI in outdoor horticulture.

The DSSs are classified into (i) model-based, (ii) sensor/measurement-based and (iii) measurement- and model-based.

Model-based DSS, such as the MODERATO management-oriented cropping system model developed by Bergez et al. (2001), simulate the plant-soil system with a dynamic biophysical model and are able to consider within-field variability. MODERATO incorporates several components that work together to achieve this simulation:

- **Hydraulic context**: This component considers factors like soil properties, water flow dynamics, and drainage characteristics.

- **Mode of action**: This component focuses on the mechanisms by which irrigation management decisions (timing, amount) influence crop growth and yield.
- **Agronomic models**: These models represent the growth and development of the specific crop being cultivated, taking into account factors like temperature, radiation, and nutrient availability.
- **Timer and agro-economic evaluator**: This component determines the optimal irrigation timing and amount based on the combined information from the other components, while also considering economic factors like water cost and crop value.

Similarly, Thorp et al. (2008) demonstrated the use of the DSSAT crop DSS model and McCarthy et al. (2010, 2014a, b) demonstrated the use of the Australian cotton model OZCOT to simulate spatial variability of crop yield and to create spatially variable PI prescription maps. In Australia, a model-based VRI-DSS (WaterSense™ https://www.watersense.eu/) was developed to help growers with water allocation restrictions. It aimed to enable growers to schedule irrigation under uncertain water supply through the season by working out the most likely yield achievable with a given allocation, soil type, planting or rotation date and past rainfall and irrigation. WaterSense compliments other work by Australian researchers who have developed automation and control methods for surface flood irrigation systems, for example Hedley et al. (2014) and Khatri and Smith (2007).

Coates et al. (2006) developed a horticulture DSS system that will allow for the management of individual trees in an orchard by using a measurement-based DSS. Customised software was used to operate the master controller, drip line controller, and individual micro-sprinkler nodes to differentially supply water and dissolved chemical fertilizers to one or more individual trees fed by a single micro-sprinkler drip line. Similarly, a dynamic VRI–DSS (http://www.myfieldnet.com/fieldnet-advisor) has recently been developed in a GIS framework to be integrated with a center pivot system to account for within-field variability of soils, often resulting in the delineation of soil management zones. This VRI–DSS manages multiple crops under one VRI system, allowing farmers to immediately activate or modify their irrigation decisions (e.g., irrigation depth) for each soil-crop combination. The VRI–DSS incorporates up-to-date information from the DTN virtual weather service, crop demand predictions, and irrigation scheduling options.

CSIRO in Australia has developed different DSS techniques for providing low-cost irrigation scheduling information over large areas using satellite and mobile phone SMS technology. Remotely sensed vegetation indices (NDVI, Ratio Vegetation Index (RVI) etc) from ground, airborne and satellite platforms have been used for determining crop coefficients and site-specific irrigation water requirements (Hornbuckle et al. 2008, 2009).

Recent opportunities provided by the Internet of Things (IoT), real-time sensor monitoring combined with big data and data analytics has improved our ability to develop technology-rich DSS for horticulture crops. For example, El-Naggar and

Hedley (2023a, b) developed a site-specific DSS to estimate ETc and deliver this information directly to the user. This DSS integrates IoT data into the FAO-56 Penman–Monteith algorithm, providing a site-specific estimate of ETc. The study was conducted at two contrasting sites in New Zealand: one for pea and bean crops in Manawatū and another for barley in Hawke's Bay. These sites were managed within two irrigation management zones using VRI systems. Similiarly, Liakos et al. (2015) developed and tested a soil moisture sensor-based VRI DSS. The DSS consisted of a WSN of soil moisture sensors installed in delineated irrigation management zones, a friendly graphical user interface (GUI) and a VRI-enabled centre pivot irrigation system. The scheduling recommendations were downloaded to the VRI controller on the centre pivot as a PI prescription map.

Furthermore, Evett et al. (2020) developed a VRI DSS "Irrigation Scheduling Supervisory Control and Data Acquisition (ISSCADA)" system using WSN sensor-based plant and soil water feedback as implemented in four U.S. states. Also, Andrade et al. (2020) discussed a comprehensive sensor-based DSS for VRI management by integrating a developed software package, ARSPivot with the ISSCADA system. Through a GUI involving a geographic information system (GIS), growers can visualize and modify VRI prescription maps that have been automatically generated using sensor-based irrigation scheduling methods, and to control and monitor the irrigation application specified in these recommended prescription maps using centre pivots.

In summary, in order to optimise WUE using PI technologies, a further layer of decision support is required. Spatio-temporal information with sufficient resolution on soil water supply and horticulture crop water demand needs to be monitored and modelled continuously through the season to determine the irrigation timings and amounts that give the desired crop response. Present proximal and remote sensor technologies (e.g. EMI, LiDAR and Vis-NIR spectral sensors) have the potential to meet these information requirements. The availability of GNSS, Geographic Information Systems (GIS), and high-speed internet connectivity make on-site or cloud-based data processing feasible within minutes. Also, recent innovations in low-power sensor and WSNs technologies combined with advances in IoT are offering tremendous opportunities for the development of DSSs with a real-time component for irrigation. Customised software based on models and sensors can improve PI for horticulture crops and water management as discussed (Sects. 6.3 and 6.4).

However, despite progress in the development of DSSs over the past decades, adoption by farmers is limited probably due, at least in part, to the complexity of technical operation and maintenance of the DSSs. In addition, systems are typically customised to specific areas, purposes and crops, so it may be difficult to apply the system to other sites.

6.5 Examples of Precision Irrigation Systems Applied in Field Horticulture, and Resulting Water, Nutrient and Yield Benefits

This section summarises a range of studies that were conducted to investigate a wide variety of site-specific irrigation management or PI strategies and their impacts on outdoor horticulture crop yields.

Annual Crops

PI management is most likely to be financially beneficial for annual horticulture crops where yield and quality are highly water sensitive and crop price is highly dependent upon crop quality. Potatoes are one of those water sensitive crops, and a field study conducted by King et al. (1999) for measuring the yield of potatoes under a VRI centre pivot showed yields per unit of water applied were greater (4 and 6%) in two consecutive years over those for uniform irrigation management.

A range of other studies have investigated the benefits of PI strategies for potato production and environmental outcomes. For instance, in New Zealand, Hedley et al. (2009) compared irrigation water use, drainage water loss, nitrogen leaching, energy use, and irrigation WUE between uniform rate irrigation and VRI at three contrasting sites using 2 years of climate data for a potato crop. The results indicated that VRI used 9–19% less irrigation water, with accompanying energy saving. Drainage volume was also reduced by 25–45% by using VRI during the period of irrigation, which reduced the risk of nitrogen leaching. Also, Hedley et al. (2010) reported potential water savings under potatoes with VRI of 8–21% while drainage and runoff were reduced 19–55% and cost savings were estimated at USD$30–95 per hectare.

Similarly, Aghil et al. (2018) investigated the use of VRI to optimise WUE and obtain high-quality potato production in Southern Alberta, Canada during the 2015 and 2016 growing seasons. Overall, 43% less water was applied under the VRI, and a 12% reduction in irrigation pumping cost was achieved.

The influence of applying PI for overhead systems as an alternative to furrow and drip irrigation was conducted by Mitchell et al. (2016). The study was conducted from 2008 through 2013, in studies in California for tomato, onion and broccoli. Overhead irrigation management was based on matching water application rates with infiltration characteristics of the soil and avoiding frequent, light irrigations which resulted in greater evaporation losses. The main findings showed equal or increased yields with best site-specific irrigation management practices for all crops except tomatoes.

Contaminant losses (nutrients and agro-chemicals) are likely to decrease with PI. This is because PI systems are able to reduce deep percolation in areas with a shallow water table, reduce application depths in areas with poor drainage, and reduce runoff in areas with poor surface drainage that may lead to poor soil aeration and consequently yield loss. Meisinger and Delgado (2002) presented principles for managing nitrate-N (NO_3-N) leaching using site-specific irrigation management. This was reported for the spatial variability of residual soil NO_3-N and NO_3-N leaching during the growing season of potatoes for centre-pivot irrigated systems. The residual soil NO_3-N in a loamy sand zone was lower than that measured for the sandy loam zone. The NO_3-N leached from the irrigated crop in the loamy sand zone areas was therefore assumed to be higher than for these crops grown on the sandy loam zone.

Long term precision agriculture research (15 years) conducted by Cambouris et al. (2014) also showed that the site-specific irrigation management approach based on a detailed soil survey map, soil properties, or, ideally, on EC_a (a numeric dataset), was the most effective approach for managing variability and enhancing potato productivity and profitability while reducing the environmental impact of production practices.

A similar positive benefit from PI application was reported by El-Naggar et al. (2020) who used VRI strategies for reduced application depths in poorly drained soils and consequently increased water productivity and irrigation WUE by 27–45% for pea and bean crop trials.

There is a growing awareness that integrated information systems monitoring soil and/or crop conditions with irrigation prescription maps used with appropriate irrigation equipment, e.g. variable-rate irrigation machines, are required for effective PI. For instance, in dry areas in Australia, a range of studies, such as Hussain (2010) and Padhi (2009), were designed to identify the spatial variability of vegetable response to PI application where low-cost satellite imagery and camera-based sensors were used to quantify crop size and vigour and to relate this to irrigation strategy.

Another study conducted by Chang and Lin (2018) developed a multi-tasking farm machine attached to an automated guided vehicle with monitoring sensors for PI. The machine can automatically weed and perform VRI within a cultivated field. The experiment was used in lettuce crop trials. The research findings highlighted the importance of monitoring the wetted distribution area of surface soil and the SWC of the deep soil to inform and drive the water pumps and perform VRI to achieve water savings.

An integrated modelling approach study conducted by Perea et al. (2018) compared the impacts of conventional and PI management strategies on crop yield and soil water management at field scale. The methodology was applied to a case study in eastern England on a high-value horticultural crop (onions). The study results indicated that VRI achieved water savings at the farm-scale due to reductions in infiltration and drainage, and the final yield increased under the VRI strategy because of higher optimised water content in the root zone of the onion crop.

More recent studies have been investigating relationships between crop yield, soluble solids content and water stress for conventional compared to variable application rates (e.g. Takács et al. 2020). In general, the research has considered variable rate application based on ET_c and crop stage as one of the most common approaches for improving WUE. In Hungary, Takács et al. (2020) designed a study for 2 years to irrigate tomatoes with a centre pivot capable of VRI. Variable water applications were 100%, 75%, and 50% of ET_c until the beginning of the ripening stage, calculated by AquaCrop, compared to a control with regular irrigation. The results suggested that deficit irrigation is very beneficial because the concentration of lycopene (a bright red carotenoid hydrocarbon found in tomatoes and other red fruits and vegetables) in the tomato fruit increases, especially in the case of the "75% of ET_c water supply" treatment until the beginning of the ripening stage. In conclusion, the "75% of ET_c water supply" treatment supplied until the beginning of ripening was a balanced water supply level regarding yield quantity, soluble solids content and lycopene concentration and yields, evidenced by higher concentration and ratio to total carotenoids.

Similarly, Takács et al. (2019) examined the yield, soluble solids content and water stress level of processing tomato—to investigate how these plant properties react to deficit irrigation (different water depths) and water distribution uniformity provided by a VRI system. The results showed that yields, soluble solids content and water stress represented the expected differences between 100% and 50% application rates. In conclusion, centre pivots with individual sprinkler control VRI systems were suitable for conducting deficit irrigation experiments concerning over- and under-irrigation and water distribution uniformity.

Perennial Crops

Most orchards are now micro-irrigated because of high efficiency compared with flood or sprinkler irrigation. Another advantage is that foliage stays dry and fungal diseases are not spread as readily. Most recent irrigation scheduling initiatives for PI of fruit trees have used plant-based scheduling methods, such as IRT sensors and heat-pulse sap flow probes (discussed in Sect. 6.3.1.2.2).

For instance, Rojo et al. (2016) conducted a research study on VRI to assess plant water stress of grapes and almond trees by comparing CWSI with deficit stem water potential. Irrigation scheduling was implemented based on monitoring Tc and microclimatic variables. Two management zones were created using unsupervised fuzzy classification based on soil and plant characteristics. In each management zone, two treatments were implemented: grower-based and stress-based. The main research findings highlighted differences between the crop water stress levels between the two management zones.

In grapes, CWSI was found to be linearly related to deficit stem water potential with a coefficient of determination value of 0.70. In almonds, CWSI was found to be strongly correlated to deficit stem water potential with a second order relationship and a coefficient of determination value of 0.78. Zone 1 required about 70% water, i.e. 30% less water compared to the grower treatment. Zone 2 required about 90% water, i.e. 10% less compared to the grower treatment.

Similarly, in Spain, Gonzalez-Dugo et al. (2020) conducted research to relate the actual transpiration of almond trees measured with heat-pulse sap flow probes with the CWSI and to assess the sources of uncertainty in the estimates of transpiration from CWSI values. The results indicated a large scatter obtained between CWSI and the relative transpiration. The study concluded the sources that may create the scatter are micrometeorological conditions in the different parts of the canopy, the portion of the tree crown targeted by infrared thermometers, and methodological errors associated with the measurement of tree transpiration and the estimation of the CWSI. All these potential factors result in inaccuracies in determining actual CWSI for the assessment of tree water status.

The relationship between tree size and tree water use has been investigated by a range of studies. These studies have shown that tree size is linearly related to tree water use. For example, Smith et al. (2011) reported VRI for trees to match ET_c resulted in maximum water use efficiency. However, such management, although used in research studies, is currently not feasible at the operational scale. In contrast, dividing the orchard into 24 row-based irrigation management units based on tree size was feasible and resulted in substantial improvements in water use efficiency. Smith et al. (2011) reported a water-saving of 1.7 ML/ha and a reduction in drainage from 4.9 to 3.2 ML/ha was estimated for this orchard.

In recent research, the application of proximal sensing and remote sensing methods for site-specific management of perennial crops has been investigated by several researchers. For example, the possibility of creating site-specific management zones within an almond orchard and a vineyard and how to determine the number and optimum locations for installing proximal sensors (leaf monitors) for determining plant water status, was investigated by Bazzi et al. (2019). Soil texture, digital elevation and EC_a (0.3 and 0.9 m soil depth) data were used to define the management zones using Fuzzy C-Means algorithm, and midday stem water potential data were used to define the best places to put a number of leaf monitor sensors within each zone. The methodology successfully informed the number of sensors needed and the location (trees) in which these sensors should be installed to better represent the data used for analysis in a grapevine vineyard and an almond orchard.

Another study conducted by Sanchez et al. (2017) used a VR-drip irrigation system to schedule irrigation in a vineyard in California based on NDVI derived from Landsat images and differentially applied management factors. The results showed increases in yield during three seasons by an average of 10% with an up to 17% gain in WUE. In comparison, Gemtos et al. (2013) used NDVI maps to identify spatial variability in yield and fruit quality for small apple orchards and vineyards. The results for both crops showed that there was high variability in the measured NDVI parameters. NDVI maps indicated that fruit quality was inferior in areas where yield was high. NDVI maps in apple orchards at the beginning of June were correlated to the final yield variability. The research findings highlight the importance of NDVI maps as an early warning of yield variability and that they can be used to manage the predicted variability.

The feasibility and effectiveness of using ER and soil moisture sensing technologies with a VR-drip irrigation system was investigated by Ortuani et al. (2019). The study was conducted in a 1-hectare productive vineyard in Northern Italy, managed during the 2018 agricultural season. ER maps were used to define irrigation management zones. Irrigation scheduling was driven using information provided by soil moisture probes installed in management zones. Results showed a reduction of water consumption by 18% compared to conventional drip management without losses in yield and product quality. Also, results for research trials conducted by Coates et al. (2006) showed that spatially variable management for VR-drip-irrigation systems is viable.

As well as defining the spatial heterogeneity of soil using PS and RS tools, Robinson et al. (2017) developed a new web-based irrigation model (Cornell Apple Irrigation Model) that uses weather data from a network of weather stations to determine the quantity of irrigation water to apply to both young and mature apple orchards of various densities. The frequency of adding the required water depends on soil type. This allows growers to more precisely manage tree water status in both wet and dry years for consistently achieving the optimum fruit size.

In conclusion, the examples discussed in this Section show the benefits of PI systems applied in outdoor horticulture. Combining crop and soil sensing information with DSS can provide the highest level of irrigation performance and the potential for spatially varied irrigation application. High resolution spatial information about soil and plant characteristics can provide a strong basis for the establishment of management zones in irrigated crops. Furthermore, WUE of crops could be improved if irrigations are scheduled when water stress reaches a certain threshold. This threshold will vary for different horticulture crops and different crop stages and depend on the frequency with which irrigation can be controlled accurately by a feedback of WSN of soil moisture sensors or Tc sensors installed in the crop root zone. Automated irrigation using feedback from these sensors makes it possible to meet the crop water requirements precisely. This produces the desired plant responses, and hence high yields, while using the exact volume of water required to maintain the crop.

6.6 Concluding Remarks

Sections 6.3 and 6.4 provided examples of the development and integration of in-field sensing systems linked to models and control interfaces that can enable precision management of irrigated horticulture and arable land. They demonstrated how the IoT with aligned technological advances have enabled significant advances over recent years in the development of automated PI systems. Section 6.5 provides examples of PI management on outdoor horticulture crops in action and presents the multiple benefits of water savings, reduced drainage and nutrient loss, often without yield penalty and sometimes with yield gain.

However, despite these advances, demonstrated largely by researchers working with the irrigation community, PI is still not widely accepted by practitioners. This is probably, in part, due to the initial cost of the required infrastructure, as well as a need for greater understanding about how to manage such systems. For example, in Australia and New Zealand, approximately 10–15% of new centre pivots have the

VRI technology installed, but it is a steep learning curve for practitioners to use the control software effectively. More consultants are needed to assist this transition. Globally, growers spoken to regarding the technology have provided a range of concerns including cost. However, in this regard Yule et al. (2013) showed payback periods ranging from 5 to 20 years for adoption of VRI in dairy and cropping in New Zealand.

Because PI is a knowledge-based management practice, there are knowledge gaps in many factors associated with PI decisions. For example, Raine et al. (2007) estimated that 10% of the irrigated land area within Australia could be adversely affected by root zone salinity resulting from the adoption of PI, which they attribute to a lack of understanding of all aspects of the whole system. Presumably, in this case, too much underwatering with insufficient drainage may cause the onset of salinity in regions that have been traditionally flood irrigated. Other factors that may contribute to the lack of adoption by farmers include the fact that PI decisions are generally made by busy people with limited technical background and training in PI management.

Emerging technical support from commercial companies for farmers and growers needs to be developed further to enable the potential water savings, yield benefits and environmental gains that are possible with PI methods. Canavari et al. (2021) recommended that irrigation professionals should clarify the benefits of PI to growers in order to increase adoption rates.

Evans et al. (2013), in the USA, indicated that the significant under-utilization of VRI systems is likely to continue into the future until (1) cost-effectiveness is increased by higher water and energy costs, (2) regulatory limits on water application amounts are implemented broadly, (3) suitable economic incentives in compliance with environmental and other regulations are implemented and enforced, and (4) information on how to manage these systems with demonstrated increased economic returns is illustrated with compelling regional research results.

There is a pressing need for the agricultural and horticultural sector to adopt PI technologies to maximize economical return per unit of water and minimise negative environmental impacts (drainage, nutrient leaching, greenhouse gas emissions, salinity). This is required due to decreasing water quality and increasing water scarcity and food security issues globally, exacerbated by climate change. PI application can maximize WUE and energy savings while improving environmental quality of irrigated cropland.

However, researchers need to work with industry and practitioners to clearly demonstrate these positive effects and quantify resulting economic and social benefits. Applied research is required to assist farmers and growers to understand how to manage VRI systems.

Variable rate fixed (solid-set) and sprinkler irrigation systems are now commercially available, but insufficient extension examples exist that conclusively demonstrate the many benefits of precision to encourage their adoption. More research is needed to develop DSS for other VRI systems, such as for travelling gun systems and drip irrigation systems.

A wide range of sensor technologies exist to map and monitor fields so that PI decisions may be implemented. A decision about which type of sensor is required will in part relate to site-specific conditions of soil, land use and climate, as well as affordability and understanding. For example, in arid regions, crop water demand can be estimated without having to take into account the effects of daily variable precipitation patterns and consequent changes in soil moisture content of different soil horizons with varying drainage characteristics. However, in humid regions, such as Canada, New Zealand and the United Kingdom, irrigation is often needed as a supplemental source of water for those times when rainfall is insufficient to meet crop water needs. Therefore, PI solutions must be customised to site-specific factors. An appropriate design is key to success, together with skilful management of the system. A long-term goal is to develop easy-to-use methods for precision control of all irrigation systems. This will take advantage of the rapid success of the IoT and advanced control strategies that the IoT enables to achieve improved monitoring and control of irrigated fields.

The control systems that are being developed rely on predictive climate forecasting and more research is needed to develop more reliable simulation tools and technologies for estimating the spatio-temporal distribution of global rainfall and climate change impacts, as well as soil water and crop requirements using satellite data integrated with verified sensing tools.

The application of artificial intelligence and machine learning approaches in PI are new principles that are beginning to be widely used by the scientific community. The adoption of these smart irrigation approaches can best improve the application of irrigation water.

Competing Interests The authors have no conflicts of interest to declare that are relevant to the content of this chapter.

References

Abuzar M, McAllister A, Morris M (2001) Classification of seasonal images for monitoring irrigated crops in a salinity-affected area of Australia. Int J Remote Sens 22(5):717–726

Aghil Y, Chandra AM, Shelley AW, Viacheslav IA, Laura G (2018) Application of variable-rate irrigation for potato productivity. Paper presented at the A paper from the Proceedings of the 14th international conference on precision agriculture, Montreal, Quebec, Canada

Alarcón J, Domingo R, Green S, Nicolás E, Torrecillas A (2003) Estimation of hydraulic conductance within field-grown apricot using sap flow measurements. Plant Soil 251(1):125–135

Andrade MA, O'Shaughnessy SA, Evett SR (2020) ARSPivot, a sensor-based decision support software for variable-rate irrigation center pivot systems: part A. Dev Trans ASABE 63(5): 1521–1533

Anguraj DK, Mandhala VN, Bhattacharyya D, Kim T-h (2021) Hybrid neural network classification for irrigation control in WSN based precision agriculture. J Ambient Intell Humaniz Comput:1–12

Barker JB, Heeren DM, Neale CM, Rudnick DR (2018) Evaluation of variable rate irrigation using a remote-sensing-based model. Agric Water Manag 203:63–74

Bazzi CL, Schenatto K, Upadhyaya S, Rojo F, Kizer E, Ko-Madden C (2019) Optimal placement of proximal sensors for precision irrigation in tree crops. Precis Agric 20(4):663–674

Bergez J-E, Debaeke P, Deumier J-M, Lacroix B, Leenhardt D, Leroy P, Wallach D (2001) MODERATO: an object-oriented decision tool for designing maize irrigation schedules. Ecol Model 137(1):43–60

Biggs T, Thenkabail P, Gumma M, Scott C, Parthasaradhi G, Turral H (2006) Irrigated area mapping in heterogeneous landscapes with MODIS time series, ground truth and census data, Krishna Basin, India. Int J Remote Sens 27(19):4245–4266

Brevik E, Fenton T, Lazari A (2006) Soil electrical conductivity as a function of soil water content and implications for soil mapping. Precis Agric 7(6):393–404

Cambouris AN, Zebarth BJ, Ziadi N, Perron I (2014) Precision agriculture in potato production. Potato Res 57(3–4):249–262

Canavari M, Medici M, Wongprawmas R, Xhakollary V, Russo S (2021) A path model of the intention to adopt variable rate irrigation in Northeast Italy. Sustain For 13(4):1879

Capraro F, Schugurensky C, Vita F, Tosetti S, Lage A, Patiño D (2008) Intelligent irrigation in grapevines: a way to obtain different wine characteristics. IFAC Proc Vol 41(2):2950–2955

Čermák J, Kučera J, Nadezhdina N (2004) Sap flow measurements with some thermodynamic methods, flow integration within trees and scaling up from sample trees to entire forest stands. Trees 18(5):529–546

Chang C-L, Lin K-M (2018) Smart agricultural machine with a computer vision-based weeding and variable-rate irrigation scheme. Robotics 7(3):38

Coates R, Delwiche M (2008) Site-specific water and chemical application by wireless valve controller network. Paper presented at the 2008 Providence, Rhode Island, June 29–July 2, 2008

Coates R, Delwiche M, Brown P (2006) Design of a system for individual microsprinkler control. Trans ASABE 49(6):1963–1970

Cohen Y, Fuchs M, Falkenflug V, Moreshet S (1988) Calibrated heat pulse method for determining water uptake in cotton. Agron J 80(3):398–402

Conaty W (2010) Temperature time thresholds for irrigation scheduling in precision application and deficit furrow irrigated cotton. PhD Thesis, Faculty of Agriculture, Food and Natural Resources, The University of Sydney, NSW, Australia

Corwin D, Lesch S (2003) Application of soil electrical conductivity to precision agriculture. Agron J 95(3):455–471

Corwin D, Lesch S (2005) Characterizing soil spatial variability with apparent soil electrical conductivity: I. Survey protocols. Comput Electron Agric 46(1):103–133

Cremona M, Stützel H, Kage H (2004) Irrigation scheduling of kohlrabi (Brassica oleracea var. gongylodes) using crop water stress index. HortScience 39(2):276–279

D'Urso G, Vuolo F, De Michele C (2008) Remote sensing techniques to improve on-farm irrigation efficiency. Paper presented at the Irrigation Australia 2008 – National Conference and Exhibition, 20–22 May, Melbourne

Daccache A, Knox JW, Weatherhead E, Daneshkhah A, Hess T (2015) Implementing precision irrigation in a humid climate – recent experiences and on-going challenges. Agric Water Manag 147:135–143

Dalla Corte AP, Souza DV, Rex FE, Sanquetta CR, Mohan M, Silva CA et al (2020) Forest inventory with high-density UAV-Lidar: machine learning approaches for predicting individual tree attributes. Comput Electron Agric 179:105815

Das I, Shah NG, Merchant SN (2010) AgriSens: wireless sensor network in precision farming: a case study. LAP Lambert Academic. ISBN: 978-3-8433-5525-4

DeJonge K, Kaleita A, Thorp K (2007) Simulating the effects of spatially variable irrigation on corn yields, costs, and revenue in Iowa. Agric Water Manag 92(1):99–109

Doerge T, Kitchen N, Lund E (1999) Site-specific management guidelines: soil electrical conductivity mapping. SSMG-30. Potash & Phosphate Institute, Atlanta

Doolittle J, Sudduth K, Kitchen N, Indorante S (1994) Estimating depths to claypans using electromagnetic induction methods. J Soil Water Conserv 49(6):572–575

Duncan H, Leib B (2011) Using GPR and soil EC to identify spatial variability of water holding capacity for precision irrigation management. Paper presented at the AGU Fall Meeting Abstracts

Ehlert D, Adamek R, Horn H-J (2009a) Vehicle based laser range finding in crops. Sensors 9(5): 3679–3694

Ehlert D, Adamek R, Horn H-J (2009b) Laser rangefinder-based measuring of crop biomass under field conditions. Precis Agric 10(5):395–408

Ehlert D, Heisig M, Adamek R (2010c) Assessment of a laser scanner on agricultural machinery. Paper presented at the Engineering for Rural Development-9th International Scientific Conference. University of Agriculture, Jelgava, Latvia

Ehlert D, Heisig M, Adamek R (2010d) Suitability of a laser rangefinder to characterize winter wheat. Precis Agric 11(6):650–663

Ehrler W (1973) Cotton leaf temperatures as related to soil water depletion and meteorological factors. Agron J 65(3):404–409

Ekanayake JC, Hedley CB (2018) Advances in information provision from wireless sensor networks for irrigated crops. Wirel Sens Netw 10(04):71

El Chami D, Knox JW, Daccache A, Weatherhead EK (2019) Assessing the financial and environmental impacts of precision irrigation in a humid climate. Hortic Sci 46(1):43–52

El-Naggar AG (2020) New sensing methods for scheduling variable rate irrigation to improve water use efficiency and reduce the environmental footprint: a thesis presented in partial fulfilment of the requirements for the degree of Doctor of Philosophy in Soil Science at Massey University, Palmerston North, New Zealand. (Doctor of Philosophy (PhD) Doctoral). Massey University, Retrieved from http://hdl.handle.net/10179/15751

El-Naggar A, Hedley C (2023a) Field evaluation of a commercial variable-rate irrigation decision support system – a study for maize and sweet corn. N Z J Agric Res. https://doi.org/10.1080/00288233.2023.2251930

El-Naggar A, Hedley C (2023b) Site-specific estimation of spatiotemporal crop evapotranspiration through field-level sensing data. Agron J. https://doi.org/10.1002/agj2.21331

El-Naggar AG, Hedley CB, Horne D, Roudier P, Clothier BE (2020) Soil sensing technology improves application of irrigation water. Agric Water Manag 228:105901. https://doi.org/10.1016/j.agwat.2019.105901

El-Naggar AG, Hedley CB, Roudier P, Horne D, Clothier BE (2021a) Imaging the electrical conductivity of the soil profile and its relationships to soil water patterns and drainage characteristics. Precis Agric. https://doi.org/10.1007/s11119-020-09763-x

El-Naggar AG, Jolly B, Hedley CB, Roudier P, Horne D, Clothier BE (2021b) The use of terrestrial LiDAR to monitor crop growth and account for within-field variability of crop coefficients and water use. Comput Electron Agric 106416. https://doi.org/10.1016/j.compag.2021.106416

Erdem Y, Arin L, Erdem T, Polat S, Deveci M, Okursoy H, Gültaş H (2010) Crop water stress index for assessing irrigation scheduling of drip irrigated broccoli (Brassica oleracea L. var. italica). Agric Water Manag 98(1):148–156

Evans R, LaRue J, Stone K, King B (2013) Adoption of site-specific variable rate sprinkler irrigation systems. Irrig Sci:1–17

Evett SR, O'Shaughnessy SA, Andrade MA, Colaizzi PD, Schwartz RC, Schomberg HS et al (2020) Theory and development of a VRI decision support system: the USDA-ARS ISSCADA approach. Trans ASABE 63(5):1507–1519

FAOSTAT data (2022) Pocketbook, FAO Statistical. "World food and agriculture". http://www.fao.org/documents/card/en/c/cb1521en/

Fernández JE, Green SR, Caspari HW, Diaz-Espejo A, Cuevas MV (2008) The use of sap flow measurements for scheduling irrigation in olive, apple and Asian pear trees and in grapevines. Plant Soil 305(1):91–104

Fraisse C, Heermann D, Duke H (1993) Modified linear move system for experimental water application. Paper presented at the Advances in planning, design and management of irrigation systems as related to sustainable land use, Leuven (Belgium), 14–17 Sep 1992

Gemtos T, Fountas S, Tagarakis A, Liakos V (2013) Precision agriculture application in fruit crops: experience in handpicked fruits. Procedia Technol 8:324–332

González-Dugo M, Mateos L (2008) Spectral vegetation indices for benchmarking water productivity of irrigated cotton and sugarbeet crops. Agric Water Manag 95(1):48–58

Gonzalez-Dugo V, Testi L, Villalobos FJ, López-Bernal A, Orgaz F, Zarco-Tejada PJ, Fereres E (2020) Empirical validation of the relationship between the crop water stress index and relative transpiration in almond trees. Agric For Meteorol 292:108128

Green S, Clothier B (1988) Water use of kiwifruit vines and apple trees by the heat-pulse technique. J Exp Bot 39(1):115–123

Grote K, Anger C, Kelly B, Hubbard S, Rubin Y (2010) Characterization of soil water content variability and soil texture using GPR groundwave techniques. J Environ Eng Geophys 15(3): 93–110

Hedley C, Yule I (2009a) A method for spatial prediction of daily soil water status for precise irrigation scheduling. Agric Water Manag 96(12):1737–1745

Hedley C, Yule I (2009b) Soil water status mapping and two variable-rate irrigation scenarios. Precis Agric 10(4):342–355

Hedley C, Yule I, Eastwood C, Shepherd T, Arnold G (2004) Rapid identification of soil textural and management zones using electromagnetic induction sensing of soils. Soil Res 42(4): 389–400

Hedley C, Yule I, Tuohy M, Vogeler I (2009) Key performance indicators for variable rate irrigation implementation on variable soils. Paper Number 096372. In: Proceedings of ASABE annual international meeting, June 21–24, 2009, Reno, Nevada

Hedley C, Yule I, Bradbury S (2010) Analysis of potential benefits of precision irrigation for variable soils at five pastoral and arable production sites in New Zealand. Paper presented at the 19th World Soil Congress

Hedley C, Ekanayake J, Roudier P (2012) Wireless sensor networks for precision irrigation scheduling. In: Currie LD, Christensen CL(eds) Advanced nutrient management: gains from the past – goals for the future. Occasional report no. 25. Fertilizer and Lime Research Centre, Massey University, Palmerston North, 8 p. http://flrc.massey.ac.nz/publications.html

Hedley C, Knox JW, Raine SR, Smith R (2014) Water: advanced irrigation technologies. In: Alfen NKV (ed) Encyclopedia of agriculture and food systems. Academic, Oxford, pp 378–406. https://doi.org/10.1016/B978-0-444-52512-3.00087-5

Hezarjaribi A, Sourell H (2007) Feasibility study of monitoring the total available water content using non-invasive electromagnetic induction-based and electrode-based soil electrical conductivity measurements. Irrig Drain 56(1):53–65

Hofle B (2014) Radiometric correction of terrestrial LiDAR point cloud data for individual maize plant detection. IEEE Geosci Remote Sens Lett 11(1):94–98

Hornbuckle J, Car N, Christen E, Smith D (2008) Large scale, low cost irrigation scheduling-making use of Satellite and ETo weather station information. Paper presented at the Irrigation Australia, Irrigation Association of Australia National Conference and Exhibition: Share the Water, Share the Benefits

Hornbuckle J, Car N, Destombes J, Smith DJ, Christen E, Goodwin I, McClymont L (2009) Measuring, mapping and communicating the effects of poor drip irigation distribution uniformity with satellite remote sensing and web based reporting tools. Presented at the irrigation and drainage conference 2009, Irrigation Australia Ltd, Swan Hill, Vic, Australia, 18–21 Oct

Hosoi F, Omasa K (2009) Estimating vertical plant area density profile and growth parameters of a wheat canopy at different growth stages using three-dimensional portable lidar imaging. ISPRS J Photogramm Remote Sens 64(2):151–158

Huang J, Pedrera-Parrilla A, Vanderlinden K, Taguas E, Gómez J, Triantafilis J (2017) Potential to map depth-specific soil organic matter content across an olive grove using quasi-2d and quasi-3d inversion of DUALEM-21 data. Catena 152:207–217

Hubbard S, Grote K, Rubin Y (2002) Mapping the volumetric soil water content of a California vineyard using high-frequency GPR ground wave data. Lead Edge 21(6):552–559

Huisman J, Hubbard S, Redman J, Annan A (2003) Measuring soil water content with ground penetrating radar. Vadose Zone J 2(4):476–491

Hussain A (2010) Quantifying the role of irrigation uniformity on lettuce production and profitability in the Lockyer valley, Queensland (PhD Dissertation, Faculty of Engineering and Surveying. Toowoomba). University of Southern Queensland

Idso S, Jackson R, Pinter P Jr, Reginato R, Hatfield J (1981) Normalizing the stress-degree-day parameter for environmental variability. Agric Meteorol 24:45–55

Irmak S, Rathje W (2014) Plant growth and yield as affected by wet soil conditions due to flooding or over-irrigation. University of Nebraska – Lincoln Extension. In

Jackson R, Idso S, Reginato R, Pinter P (1981) Canopy temperature as a crop water stress indicator. Water Resour Res 17(4):1133–1138

Jiménez-Buendía M, Soto-Valles F, Blaya-Ros PJ, Toledo-Moreo A, Domingo-Miguel R, Torres-Sánchez R (2021) High-density Wi-Fi based sensor network for efficient irrigation management in precision agriculture. Appl Sci 11(4):1628

Jones H (2004) Application of thermal imaging and infrared sensing in plant physiology and ecophysiology. Adv Bot Res 41:107–163

Jones H (2006) Irrigation scheduling – comparison of soil, plant and atmosphere monitoring approaches. Paper presented at the V International symposium on irrigation of horticultural crops, 792

Khatri KL, Smith R (2007) Toward a simple real-time control system for efficient management of furrow irrigation. Irrig Drain 56(4):463–475

Kim Y, Evans R, Iversen W (2008) Remote sensing and control of an irrigation system using a distributed wireless sensor network. IEEE Trans Instrum Meas 57(7):1379–1387

King B, McCann I, Eberlein C, Stark J (1999) Computer control system for spatially varied water and chemical application studies with continuous-move irrigation systems. Comput Electron Agric 24(3):177–194

Kogan F, Gitelson A, Zakarin E, Spivak L, Lebed L (2003) AVHRR-based spectral vegetation index for quantitative assessment of vegetation state and productivity. Photogramm Eng Remote Sens 69(8):899–906

Köksal E (2008) Irrigation water management with water deficit index calculated based on oblique viewed surface temperature. Irrig Sci 27(1):41–56

Lane AL (2016) The use of ground penetrating radar to map soil physical properties that control water flow pathways in alluvial soils: a thesis presented in partial fulfilment of the requirement for the degree of Master of Science in Agriculture at Massey University, Manawatu, New Zealand. Massey University

Liakos V, Vellidis G, Tucker M, Lowrance C, Liang X (2015) A decision support tool for managing precision irrigation with center pivots. In: Precision agriculture'15. Wageningen Academic, pp 713–720

Lo T, Mateos L, Heeren D, Luck J (2014) The applicability of VRI for managing variability in infiltration capacity and plant-available water: A preliminary discussion and GIS study. Paper presented at the 2014 Montreal, Quebec Canada July 13–July 16, 2014

Lubczynski M, Roy J (2004) Magnetic resonance sounding: new method for ground water assessment. Groundwater 42(2):291–309

McCarthy A, Hancock N, Raine S (2010) VARIwise: a general-purpose adaptive control simulation framework for spatially and temporally varied irrigation at sub-field scale. Comput Electron Agric 70(1):117–128

McCarthy A, Hancock N, Raine S (2014a) Simulation of irrigation control strategies for cotton using model predictive control within the VARIwise simulation framework. Comput Electron Agric 101:135–147

McCarthy A, Hancock N, Raine S (2014b) Development and simulation of sensor-based irrigation control strategies for cotton using the VARIwise simulation framework. Comput Electron Agric 101:148–162

Meisinger J, Delgado J (2002) Principles for managing nitrogen leaching. J Soil Water Conserv 57(6):485–498

Mertzanides Y, Tsakmakis I, Kargiotis E, Sylaios G (2020) Electrical resistivity tomography for spatiotemporal variations of soil moisture in a precision irrigation experiment. Int Agrophys 34(3):309–319

Mitchell J, Shrestha A, Hollingsworth J, Munk D, Hembree K, Turini T (2016) Precision overhead irrigation is suitable for several Central Valley crops. Calif Agric 70(2):62–70

Nijbroek R, Hoogenboom G, Jones J (2003) Optimizing irrigation management for a spatially variable soybean field. Agric Syst 76(1):359–377

O'Shaughnessy S, Evett S (2008) Integration of wireless sensor networks into moving irrigation systems for automatic irrigation scheduling. Paper presented at the 2008 Providence, Rhode Island, June 29–July 2, 2008

O'Shaughnessy S, Evett S (2010) Developing wireless sensor networks for monitoring crop canopy temperature using a moving sprinkler system as a platform. Appl Eng Agric 26(2):331–341

Ooi SK, Mareels I, Cooley N, Dunn G, Thoms G (2008) A systems engineering approach to viticulture on-farm irrigation. IFAC Proc Vol 41(2):9569–9574

Ortuani B, Facchi A, Mayer A, Bianchi D, Bianchi A, Brancadoro L (2019) Assessing the effectiveness of variable-rate drip irrigation on water use efficiency in a Vineyard In Northern Italy. Water 11(10):1964

Ozdogan M, Woodcock C, Salvucci G, Demir H (2006) Changes in summer irrigated crop area and water use in Southeastern Turkey from 1993 to 2002: implications for current and future water resources. Water Resour Manag 20(3):467–488

Ozdogan M, Yang Y, Allez G, Cervantes C (2010) Remote sensing of irrigated agriculture: opportunities and challenges. Remote Sens 2(9):2274–2304

Padhi J (2009) Incorporating spatial variability in soil and crop properties for effective irrigation. (PhD Dissertation, Faculty of Engineering and Surveying). University of Southern Queensland,

Perea RG, Daccache A, Díaz JR, Poyato EC, Knox JW (2018) Modelling impacts of precision irrigation on crop yield and in-field water management. Precis Agric 19(3):497–512

Peters R, Evett S (2008) Automation of a center pivot using the temperature-time-threshold method of irrigation scheduling. J Irrig Drain Eng 134(3):286–291

Plant R, Keeley M (1999) Relationships among final plant map indices in Acala cotton. J Prod Agric 12(1):61–68

Raine SR, Meyer WS, Rassam DW, Hutson JL, Cook FJ (2007) Soil–water and solute movement under precision irrigation: knowledge gaps for managing sustainable root zones. Irrigation Science 26:91–100

Robinson T, Lakso A, Lordan J, Francescatto P, Dragoni D, DeGaetano A, Eggleston K (2017) Precision irrigation management of apple with an apple-specific Penman-Monteith model. Paper presented at the VIII International symposium on irrigation of horticultural crops, 1150

Rojo F, Kizer E, Upadhyaya S, Ozmen S, Ko-Madden C, Zhang Q (2016) A leaf monitoring system for continuous measurement of plant water status to assist in precision irrigation in grape and almond crops. IFAC-PapersOnLine 49(16):209–215

Rossel RAV, McBratney AB, Minasny B (2010) Proximal soil sensing. Springer

Sadler E, Camp C, Evans D, Usrey L (1996) A site-specific center pivot irrigation system for highly-variable coastal plain soils. In: Precision Agriculture (precisionagricu3), pp 827–834

Sadler E, Evans R, Buchleiter G, King B, Camp C (2000) Design considerations for site specific irrigation. Paper presented at the National irrigation symposium. Proceedings of the 4th decennial symposium, Phoenix, Arizona, USA, November 14–16, 2000

Sadler E, Evans R, Stone K, Camp C (2005) Opportunities for conservation with precision irrigation. J Soil Water Conserv 60(6):371–378

Saeys W, Lenaerts B, Craessaerts G, De Baerdemaeker J (2009) Estimation of the crop density of small grains using LiDAR sensors. Biosyst Eng 102(1):22–30

Samouëlian A, Cousin I, Tabbagh A, Bruand A, Richard G (2005) Electrical resistivity survey in soil science: a review. Soil Tillage Res 83(2):173–193

Sanchez L, Sams B, Alsina M, Hinds N, Klein L, Dokoozlian N (2017) Improving vineyard water use efficiency and yield with variable rate irrigation in California. Adv Anim Biosci 8(2): 574–577

Scherrer D, Bader M-F, Körner C (2011) Drought-sensitivity ranking of deciduous tree species based on thermal imaging of forest canopies. Agric For Meteorol 151(12):1632–1640
Singh R, Roy S, Kogan F (2003) Vegetation and temperature condition indices from NOAA AVHRR data for drought monitoring over India. Int J Remote Sens 24(22):4393–4402
Smith R, Baillie C, McCarthy A, Raine S, Baillie J (2011) Review of precision irrigation technologies and their application. National Centre for Engineering in Agriculture. Publication 1003017/1, USQ, Toowoomba
Sudduth K, Drummond S, Kitchen N (2001) Accuracy issues in electromagnetic induction sensing of soil electrical conductivity for precision agriculture. Comput Electron Agric 31(3):239–264
Sudduth K, Kitchen N, Wiebold W, Batchelor W, Bollero G, Bullock D et al (2005) Relating apparent electrical conductivity to soil properties across the north-Central USA. Comput Electron Agric 46(1):263–283
Takács S, Bíró T, Helyes L, Pék Z (2019) Variable rate precision irrigation technology for deficit irrigation of processing tomato. Irrig Drain 68(2):234–244
Takács S, Pék Z, Csányi D, Daood HG, Szuvandzsiev P, Palotás G, Helyes L (2020) Influence of water stress levels on the yield and lycopene content of tomato. Water 12(8):2165
Thorp KR, DeJonge KC, Kaleita AL, Batchelor WD, Paz JO (2008) Methodology for the use of DSSAT models for precision agriculture decision support. Comput Electron Agric 64(2): 276–285
Tsoulias N, Paraforos D, Fountas S, Zude-Sasse M (2019) Calculating the water deficit spatially using LiDAR laser scanner in an apple orchard. Paper presented at the Proceedings of the European 12th conference of precision agriculture, Montpellier, France
Wall R, King B, McCann I (1996) Center-pivot irrigation system control and data communications network for real-time variable water application. In: Precision agriculture(precisionagricu3), pp 757–766
Wang D, Gartung J (2010) Infrared canopy temperature of early-ripening peach trees under postharvest deficit irrigation. Agric Water Manag 97(11):1787–1794
White SC Raine SR (2008) A grower guide to plant based sensing for irrigation scheduling. Project Report. University of Southern Queensland, Toowoomba, Australia (Report)
Widmoser P (2010) An alternative to define canopy surface temperature bounds. Agric Water Manag 97(2):224–230
Wiegand C, Richardson A, Escobar D, Gerbermann A (1991) Vegetation indices in crop assessments. Remote Sens Environ 35(2–3):105–119
Williams B, Hoey D (1987) The use of electromagnetic induction to detect the spatial variability of the salt and clay contents of soils. Soil Res 25(1):21–27
Wong M, Harper R (1999) Use of on-ground gamma-ray spectrometry to measure plant-available potassium and other topsoil attributes. Soil Res 37(2):267–278
Yildirim M, Demirel K, Bahar E (2012) Effect of restricted water supply and stress development on growth of bell pepper (Capsicum Annuum L.) under drought conditions. J Agro Crop Sci 3(1): 1–9
Yule I, Lawrence H, Mackenzie C, Hedley C, Grafton M, Pullanagari R (2013) Case studies which demonstrate the financial viability of precision dairy farming. Magnesium 9(6):1–16
Zhang L, Grift T (2012) A LIDAR-based crop height measurement system for Miscanthus giganteus. Comput Electron Agric 85:70–76

Chapter 7
Precision Management of Crop Nutrition

Lizzie Sagoo and Paul Newell Price

Abstract Precision farming has the potential to improve nutrient use efficiency by using information on soil and crop variability to better target nutrient inputs. This chapter outlines the use of precision horticulture technologies that can be used to map soil or crop properties and used as a basis for variable rate fertiliser application. Precision soil sampling can be used to map soil phosphorus and potassium levels. Typically, one sample per hectare is taken using a grid or zone-based sampling approach. Soil phosphorus and potassium maps can be converted into prescription maps for variable rate phosphorus and potassium fertiliser application. Soil pH can be mapped with a greater level of detail using commercially available in field on-the-go sensors which measure the soil pH from direct soil contact with ion selective antimony electrodes. These soil pH maps can be converted into prescription maps for variable rate lime application. In contrast to pH, phosphorus and potassium, precision management of nitrogen is based on measuring the crop rather than the soil. Crop sensing can provide valuable information on the performance of the crop across the entire planted area (spatial) and throughout the growing season (temporal) and can be used as the basis for variable rate nitrogen management. Potential benefits include increased crop yields and quality from applying the correct amount of fertiliser or lime to areas which would otherwise have been under fertilised, savings in fertiliser or lime from not overapplying to areas with high soil nutrient/pH levels, and the potential environmental benefits associated with this. Whilst the technology is well developed and commercially available, uptake has been limited by the lack of evidence to demonstrate clear and consistent yield and economic benefits.

Keywords Crop nutrition · Crop sensing · Nutrient use efficiency · Precision nutrition · Variable rate fertiliser

L. Sagoo (✉)
ADAS Boxworth, Cambridge, UK
e-mail: Lizzie.Sagoo@adas.co.uk

P. Newell Price
ADAS Gleadthorpe, Nottinghamshire, UK

P. Newell Price et al. (eds.), *Precision Agriculture in Field Horticulture*, Progress in Precision Agriculture, https://doi.org/10.1007/978-3-031-96534-0_7

7.1 Introduction

Good nutrient management is essential to optimise crop yields and quality, and to protect the environment. Crops will obtain some, sometimes all, of their nutrient requirements from the soil; however, when supply from the soil is insufficient to meet crop requirements, crop growth can be limited, and additional nutrients in the form of fertilisers or organic materials should be added to meet crop demand.

Crops require a range of nutrient elements to grow, including the macro-nutrients nitrogen, phosphorus, potassium, calcium, magnesium and sulphur, and micro-nutrients iron, copper, manganese, zinc, boron, molybdenum and chlorine. A deficiency in any one of these nutrients can limit crop growth and compromise yields and quality. However, nitrogen, phosphorus and potassium are the most frequent yield limiting nutrients and are applied in the largest amount; in 2022 the global use of manufactured nitrogen, phosphorus and potassium fertiliser was 185 million tonnes (FAO 2024).

Whilst it is important to apply sufficient nutrients to meet crop demand and achieve optimal yields and quality, it is also important to not exceed crop nutrient demand. Over-application of nutrients, in particular nitrogen and phosphorus, can lead to air and water pollution and biodiversity loss. The planetary boundary for flows of nitrogen and phosphorus have been breached (Caesar et al. 2024).

Nitrogen can be lost to the environment via nitrate leaching, ammonia volatilisation, and nitrous oxide emissions. Nitrate leaching to ground and surface waters affects drinking water quality and can contribute to eutrophication of surface waters. Ammonia emissions can contribute to formation of fine particulate matter in the atmosphere reducing air quality and impacting on human health, and when deposited on land ammonia can contribute to acidification and eutrophication of sensitive habitats. Nitrous oxide is a greenhouse gas with a global warming potential 273 times that of carbon dioxide. Phosphorus can be lost to water as soluble phosphates or bound to particulate matter in surface water runoff, contributing to eutrophication. Both nitrogen and phosphorous can trigger dead zones in freshwater and marine ecosystems.

It is estimated that only around 60% of nitrogen (Leip et al. 2011) and 10–25% of phosphorus (Syers et al. 2008) fertiliser applied to agricultural land is taken up by crops, while the remainder either remains in the soil or is lost to the environment. Increasing nutrient use efficiency (the proportion of applied nutrients that are taken up by the crop) will help reduce nutrient losses to the environment, increase crop yields and quality, and reduce the total quantity of fertiliser required.

Many countries now have targets or legislation to reduce nutrient losses from agriculture and horticulture. For example, the European Farm to Fork strategy aims to achieve a reduction in nutrient losses from both organic and mineral fertilisers of at least 50% by 2030, while ensuring no deterioration in soil fertility, which is expected to lead to a reduction in fertiliser use of at least 20%.

Crop nutrient demand will depend on nutrient supply from the soil and crop yield potential, both of which can be spatially variable depending on soil conditions.

Conventional crop nutrient and fertiliser application provides a uniform application of fertiliser or lime and does not account for within field variability. This can lead to over-application of nutrients to some parts of a field and under application in others. Precision farming has the potential to improve nutrient use efficiency by using information on soil and crop variability to better target nutrient inputs.

7.2 Precision Management of Phosphorus and Potassium

Introduction

Many fields exhibit some level of spatial variation in concentrations of plant available phosphorus and potassium in the soil, and where this is the case a uniform application of phosphorus or potassium fertiliser may result in over application of fertiliser to some areas and under application to others. Within field phosphorus and potassium variability is often driven by variation in soil type and texture. As highlighted in Chap. 2 (Soil sensing and sampling), 'clayey' soils (by definition) contain a high proportion of clay mineral particles with a negative charge that contributes to cation exchange capacity; these soils can hold on to and in some cases supply potassium through weathering, but their high sorption capacity means that phosphorus availability can be naturally low. By contrast, sandy soils have a low clay mineral content and tend to have low cation exchange capacity and phosphorus sorption capacity. Their ability to hold on to and supply potassium is therefore often low, while phosphorus availability can be higher than in soils with higher clay content. However, other factors can also account for within field phosphorus and potassium variability, such as differences in field management and crop phosphorus and potassium offtake. If fields are cropped with more than one crop, as is often the case in intensive vegetable production, differences in phosphorus and potassium inputs and crop offtakes to the areas of different crops can cause variation in soil phosphorus and potassium. Even within fields with uniform crop management, spatial variation in yield can cause differential phosphorus and potassium crop offtake and contribute to variability. For example, higher yields from heavier textured, less drought prone areas will over time result in higher crop phosphorus and potassium offtake, which may reduce soil phosphorus and potassium levels, especially if phosphorus and potassium fertiliser is applied at a uniform rate based on average field crop offtake figures.

In most cases, targeted (variable rate) phosphorus and potassium fertiliser applications are based on maps produced from precision soil sampling and analysis for phosphorus and potassium. There are two main approaches to precision soil sampling—grid based, and zone based. There is an inevitable compromise with precision soil sampling between the number of soil samples taken and the cost of sampling and analysis. A higher intensity of sampling will result in a more accurate map of soil phosphorus and potassium on which to base variable rate applications, however the cost of sampling and laboratory analysis also increases. Most commercial precision farming companies offering this as a service will map soil phosphorus and potassium levels based on analysis of one sample per hectare. Research effort is being devoted to technologies that could allow more rapid in field on-the-go sensing of these nutrients and which would provide higher resolution soil nutrient maps. The most promising technologies have focussed on ion-selective electrodes for potassium, and reflectance-based measurements for phosphorus. However, challenges with accuracy and using these sensors 'on-the-go' have so far limited their use commercially.

Precision Soil Sampling and Variable Rate Phosphorus and Potassium Application

Precision soil sampling can be used to map soil phosphorus and potassium levels. A regular or grid-based soil sampling method typically takes one sample per hectare from approximately equal-sized blocks within a field. Each point is GNSS (Global Navigation Satellite System) logged, and several subsamples are taken to form a composite sample from each point—typically 16 subsamples from a 3 m radius around the GNSS logged point. Soil sample results from GNSS logged point samples can be used to create a map of phosphorus and potassium levels using a method of data interpolation such as inverse distance weight, nearest neighbour or kriging to estimate values between the measured points. The map that is produced will depend on the grid location (and therefore sampling points) selected and the data interpolation method used (see Chap. 2—Soil sensing and sampling).

Zone-based soil sampling uses existing knowledge of within-field soil variability to direct where samples are taken. Soil zones can be based on measured soil and/or crop variability, for example using soil Electrical Conductivity (EC) maps, satellite NDVI imagery, or crop yield maps. Once the zones have been defined, each is sampled separately. The soil phosphorus and potassium maps produced will reflect the boundaries between the soil zones. Case study 2.2 (Mapping soil variability) in Chap. 2 provides an example of mapping soil phosphorus and potassium levels using both grid and zone-based soil sampling.

Soil phosphorus and potassium maps can be converted into prescription maps for variable rate phosphorus and potassium fertiliser application. A prescription map is an electronic data file which is used to control a variable rate fertiliser spreader. Variable rate fertiliser application maps are typically based on local or national fertiliser recommendations for different soil phosphorus and potassium levels.

Potential advantages of variable rate phosphorus and potassium application include cost savings in phosphorus and potassium fertiliser through not over applying to areas of higher soil phosphorus and potassium levels; potential for increased yields where lower phosphorus and potassium areas of a field would otherwise have been under-fertilised; and the longer-term reduction of within-field soil phosphorus and potassium variability.

Precision soil sampling and mapping of soil phosphorus and potassium is available as a commercial service in many countries. For example, SOYL in the UK will sample, map and provide variable rate fertiliser recommendations for phosphorus and potassium (https://www.soyl.com/services/nutrient-management).

Shah et al. (2017) measured the impact of variable rate phosphorus and potassium application on soil phosphorus and potassium levels between 2013 and 2016 in six fields in Kent, UK. Soil EC mapping was used to create soil management zones, which were sampled separately for soil phosphorus and potassium, and used as the basis for variable rate fertiliser prescription maps. After 4 years of variable rate application, the variability in soil phosphorus levels was reduced in five of the six fields, and the variability in soil potassium levels was reduced in four of the six fields. Similar results were observed in an earlier study by Kulczycki and Grocholski (2013) who reported reduced soil phosphorus and potassium variability in response to targeted variable rate application of these nutrients between 2007 and 2012 to a field in Poland. However, neither study attempted to measure a yield benefit of variable rate fertiliser application.

Variable Rate Phosphorus and Potassium Application Based on Crop Nutrient Offtake

Repeated removal of phosphorus and potassium in harvested crops will reduce soil available phosphorus and potassium levels over a period of years unless the nutrients that are removed in the harvested crop are replaced by organic manure or fertiliser. In most countries, phosphorus and potassium fertiliser recommendations account for crop offtake. For example in England fertiliser phosphorus and potassium recommendations (AHDB 2023) are based on the principle of maintaining soils at 'target' phosphorus and potassium indices. Fertiliser phosphorus and potassium recommendations aim to replace crop phosphorus and potassium offtake at the target index and include an 'index adjustment' where the soil is below target index to raise the soil phosphorus and potassium Index to target over a number of years. Where the soil index is above the target index, recommendations are reduced or omitted to allow the soil index to reduce to the target index.

Yield variability across a field can result in differences in phosphorus and potassium crop offtake, and where this pattern of yield variability persists over several years, the cumulative differences in phosphorus and potassium offtake can contribute to variation in soil phosphorus and potassium levels if fertiliser phosphorus and potassium are applied based on 'average' crop phosphorus and potassium offtake figures.

If a yield map is available, this can be combined with typical figures for crop phosphorus and potassium content to produce a map of phosphorus and potassium crop offtake, which can in turn be used to inform variable rate phosphorus and potassium application. If both a soil nutrient map and crop nutrient offtake map are available, this data can be combined to produce a prescription map for variable rate phosphorus and potassium application based on soil nutrient levels and crop offtake (Heege 2013).

Crop nutrient offtake maps require information on both yield and crop nutrient content. Yield mapping is most commonly used in combinable cropping systems. Yield monitors have been available since the early 1990s, and are now routinely fitted with GPS (Global Positioning Systems) on many combine harvesters. Yield monitoring of horticultural crops is however less common. Technology for yield mapping horticultural crops exists (Longchamps et al. 2022), however its uptake is still limited. Where yield maps are available, growers also need information on crop phosphorus and potassium nutrient content to calculate crop offtake. Typical figures for harvested crop phosphorus and potassium contents for horticultural crops are not as easily accessible or based on the same number of samples as figures for broadacre combinable crops. The requirements for both yield maps and crop nutrient offtake figures mean that whilst mapping crop offtake is technically feasible, the authors are not aware of any horticultural growers integrating crop nutrient offtake maps into their phosphorus and potassium fertiliser planning. However, it should be remembered that horticultural crops are often grown in rotation with combinable crops, and it may therefore still be possible to account for variation in phosphorus and potassium offtakes for some (if not all) the crops in the rotation.

Soil Sensors

Soil sensors are instruments used to measure soil properties, usually in the field. Soil sensors have been developed to measure several different soil properties including soil moisture, pH, nutrients, texture, and compaction, using a range of different technologies. Soil nutrient sensors offer two main advantages compared to conventional wet chemistry laboratory analysis; the results from a sensor are available quicker than a laboratory analysis, often immediately in the field, and the results from sensors are often cheaper than the cost of a laboratory analysis. The main challenge in developing soil nutrient sensors has been to develop sensors that provide accurate results which are representative of a given area of land and valid under a range of soil and climatic conditions.

Most soil nutrient sensors provide a value for a specific nutrient at a single sampling point, based on a calibration database between the soil property which the sensor measures (for example, an electrochemical signal or optical reflectance measurement) and nutrient values measured by laboratory analysis. For the sensor to be of use, the calibration database must show it to provide an accurate estimate of the specific soil nutrient, and must include data from a wide range of different soil types and climatic conditions to ensure results are valid. There must also be sufficient replication, as is the case for physical soil sampling and wet chemistry analysis, to ensure that the values produced are representative of the field or area being sampled.

PES Technologies (www.pestechnologies.com) have recently launched a hand-held soil sensor to the UK market that can be used on composite soil samples, thereby enabling sufficient replication through an appropriate sampling protocol. The PES unit uses printed electronic sensors to measure volatile organic compounds (VOCs) which are calibrated to provide a measure of a range of soil properties including extractable phosphorus, potassium and magnesium, pH, extractable nitrate and ammonium, moisture content, sand, silt and clay content, microbial biomass and respiration. The sensor has been calibrated based on a database of UK soil samples and is at least 80% accurate for all parameters for which a result is provided. A small sample of soil is placed into the sample vessel in the unit, and a result is available within 5 min. The PES unit uses cassettes with 50 or 100 sensor rolls, which are cheaper than conventional laboratory analysis for the same suite of variables but may not be cheaper if the grower was only interested in one or two soil properties. The sensor will record the GPS location of the sample but will not map the analysis results. In addition to providing rapid analysis results, a key advantage of these types of in field sensors could be that they allow growers to spot check different areas of their fields, which can help improve their understanding of how soil factors affect crop growth.

Although portable soil sensors are useful and provide advantages over conventional laboratory analysis, greater benefits can be achieved from the development of rapid, low cost, accurate on-the-go soil nutrient sensors that provide higher resolution soil nutrient maps. However, there are significant challenges in translating stationary sensors to continuous or on-the-go measurements (Heege 2013). Most on-the-go soil nutrient sensors described in the literature are based on either ion-selective electrodes or spectral reflectance.

An ion-selective electrode transforms the activity of a specific ion dissolved in a solution into an electrical potential which can be used for detecting specific ions. Ion-selective electrodes have been developed to measure a soil water mix, and in order to use these sensors on-the-go, the sensing operation must either include preparation of a soil water mixture, or rely on sensing naturally moist soil. Adamchuck et al. (2005) evaluated a method of direct soil measurement using ion-selective electrodes for on-the-go measurement of soil pH, extractable potassium, nitrate-nitrogen, and sodium. The coefficients of determination (R^2) of regressions between average direct soil measurements and corresponding reference measurements were 0.93–0.96 for soil pH, 0.61–0.62 for extractable potassium, 0.41–0.51 for nitrate-nitrogen, and 0.11 for sodium. The use of ion-selective electrodes for measurement of soil pH has been developed commercially and is discussed further in Sect. 7.3. Adamchuck et al. (2005) attributed the lower corrections for potassium, nitrate-nitrogen and sodium to: (i) differences between ion activity in naturally moist soil and ion concentration in water-based extracts, (ii) ion mobility and potential change due to overnight storage of samples, and (iii) the relatively low range of ion activity represented in the data set of soils used in the study. Note that measurement of soil phosphorus using ion-selective electrodes was not included in the study by Adamchuck et al. (2005), and is generally not considered successful (Heege 2013).

Table 7.1 Correlations between optical reflectance and various soil properties from spectroscopic experiments in laboratories (from Heege 2013)

	Coefficient of determination (r^2)		
Soil property (extractants in brackets)	Chang et al. (2001)	He et al. (2007)	Lee et al. (2009)
pH (water)	0.55	0.87	0.68
Ca (ammonium acetate)	0.75	–	0.80
Magnesium (ammonium acetate)	0.68	–	0.73
Potassium (ammonium acetate)	0.55	0.58	0.13
Phosphorus (Mehlich, Olsen)	0.40	0.29	–
Copper (Mehlich)	0.25	–	–
Manganese (Mehlich)	0.70	–	–
Zinc (Mehlich)	0.44	–	–

Another technology that has received increased attention for measuring soil properties, including nutrients, is spectral reflectance. Spectral reflectance sensors provide a rapid, low-cost measure of soil properties, and can be adapted to on-the-go soil measurements (for example Maleki et al. 2008). Spectra are calibrated to estimate concentrations of soil properties with known error. Heege (2013) collated correlations between optical reflectance and various soil properties from spectroscopic experiments in laboratories reported in the literature (Table 7.1). Although these correlations indicate some potential for use of spectral reflectance to measure soil nutrient content, they are not accurate enough to use as a replacement for laboratory analysis. McBride (2022) recently cautioned the use of reflectance measurements, concluding that for real world soil testing for agronomic and environmental purposes (e.g. extractable nutrient and trace metal levels), spectral reflectance methods are not sufficiently reliable to replace conventional laboratory analysis. However, McBride (2022) also noted that spectral reflectance did show potential benefit for measuring general trends across landscapes in levels of soil organic matter, carbonates, soluble salts and specific clay minerals.

Uptake and Outlook

Variable rate phosphorus and potassium applications based on precision soil sampling and analysis is well established and available commercially in many countries. However, there is a lack of evidence demonstrating a yield or crop quality benefit from variable rate phosphorus and potassium application, and the main benefit promoted to growers is through fertiliser saving from reducing over application of fertiliser. This was highlighted by Mallarino (1998) who attributed the lack of yield response to variable rate phosphorus application in four Iowa trials with corn and soybeans to the soil phosphorus levels that were either optimum or high and therefore there was little yield response to applied phosphorus, regardless of whether it was applied uniformly or variably. The author noted that more than half of Iowa fields tested optimum or above in phosphorus and potassium, and given the likelihood of small responsive areas in many Iowa fields, the most likely benefit of precision soil sampling and variable-rate fertiliser application will be through savings in maintenance fertilisation. Without precision soil sampling, uncertainty usually leads growers to apply a uniform high maintenance rate over the field when it may not be needed.

A 2017 review of precision farming in horticulture in the UK (Sagoo et al. 2017) found contrasting views from horticultural growers on the value of variable rate phosphorus and potassium applications; some growers have adopted variable rate applications and were happy with the approach, whilst others were unconvinced of the potential benefits. Both growers and precision farming companies noted that small savings in fertiliser were less important to horticultural growers than to broadacre arable and grassland growers. However, several growers and precision farming companies highlighted the value of soil nutrient maps in their own right (not necessarily as a basis for variable rate fertiliser application), particularly for rented land where the grower may not have the field history and soil nutrient maps enabling them to better understand the land they are renting.

A more recent telephone survey of UK potato and horticultural growers (Sagoo et al. 2024) found that a third (11 of 33 growers surveyed) had used variable rate phosphorus and potassium on their crops. Although the majority (21 of 33 growers) said they knew how to use variable rate phosphorus and potassium as a management strategy, only 11 growers considered that variable rate phosphorus and/or potassium application would be effective at increasing phosphorus and/or potassium use efficiency for the crop in question; 17 growers did not think it would increase phosphorus and/or potassium use efficiency; and six growers were unsure.

7.3 Precision Management of Soil pH

Introduction

Soil pH is a measure of the acidity or alkalinity of the soil and significantly impacts plant growth by affecting the availability of essential plant nutrients. The optimum availability of most plant nutrients occurs around a neutral pH of 6.5–7, however the optimum pH for soils will vary depending on the soil type and the cropping. Soil pH is determined by soil type, climate, topography (through its influence on surface and subsurface runoff), and anthropogenic activity, notably application of fertilisers which acidify the soil and application of lime to increase soil pH. Soils with high clay content will have a higher buffering capacity than soils with little clay, and soils with high organic matter will have a higher buffering capacity than those with low organic matter. Soils with a low buffering capacity are more susceptible to acidification and will require more frequent application of lime to maintain the target pH. Spatial variation in soil type and topography can result in sometimes significant within field variability in soil pH. Case study 2.2 (Mapping soil variability) in

Chap. 2 showed soil pH values ranging from 5.3 to 7.1 across a 10-hectare arable/horticultural field in Bedfordshire, England.

In the same way as for phosphorus and potassium, where variation in soil pH can be mapped across a field, the map can be used as a basis for variable rate lime application. Soil pH can be mapped by soil sampling and analysis, and it is common for precision soil sampling for phosphorus and potassium to also include analysis and mapping of soil pH. But of greater interest is the use of soil sensors for on-the-go soil pH mapping which provides higher spatial resolution soil pH maps. The challenge of fast accurate on-the-go measurement of soil pH has been achieved unlike for soil phosphorus and potassium, and was commercialised over two decades ago with the introduction of the Veris Mobile Sensor Platform (Adamchuk et al. 2004).

On-the-Go Measurement of Soil pH

Veris® Technologies have developed two types of machines for soil pH mapping; the MSP-3 machine that can be mounted on the back of a tractor or towed by a four-wheel drive and the U3 series that can be towed by an ATV. As these machines are towed across the field, they take a soil sample on-the-go, measure the soil pH from direct soil contact with ion selective antimony electrodes, and record the geographic position. After each sample the electrodes are washed to avoid contamination between one point and the next. This process is repeated every 20–30 s. The data is converted to the equivalent pH measured in $CaCl_2$ and a pH map produced. Early research work with the Veris® sensors demonstrated good correlations between pH measured by the Veris® sensors and pH measured by laboratory analysis, with R^2 values of 0.93–0.96 reported by Adamchuck et al. (2005). With a travel speed of 2 m/s (7.2 km/h) and one measurement every 25 s, eight readings are obtained per hectare. The sampling intensity can be increased or decreased by varying the forward speed or sample settings on the instrument.

Update and Outlook

Soil pH mapping and variable rate lime application will be of particular benefit in fields with areas of lower than expected pH, and allows correction of this through targeted lime applications.

Grower demonstration work in Queensland, Australia reported by Limpus et al. (2016) demonstrated the potential yield benefits from mapping soil pH and variable rate lime applications in intensive vegetable production systems. Strategic soil sampling of a pivot quarter (area of production based on pivot irrigation) detected areas of low pH which were mapped and used to produce a prescription map for variable rate lime application. Lime rates were assigned based on a targeted pH increase from 5.5 to 6.5. Traditionally, a fixed rate of lime (approximately 1–3 t/ha) would have been applied across the whole field. However, using variable rate application, up to 6.5 t/ha of lime was applied to the areas of lower pH and a reduced rate of 3 t/ha was applied to the remaining area. The case study did not result in a saving in lime or cost of spreading; however, it did result in noticeable yield improvements. In the subsequent green bean crop, there was a 75% yield response across the lower pH areas, which equated to approximately 28% of the field. The variable rate lime prescription used 14% more lime, but it was applied in a more targeted manner and achieved a yield result which more than offset the cost of the additional lime.

Targeted 'precision' soil sampling for pH represents an improvement in accuracy compared to 'whole field' soil sampling. However, for soil pH, further improvements can be made through on-the-go soil pH measurements. The higher sampling intensity provided by on-the-go soil pH mapping can provide more detailed information on soil pH and in turn enable more accurate variable rate lime recommendations, compared to conventional precision soil sampling with one sample per hectare. Lund et al. (2004) compared variable rate lime prescription maps produced for five fields using the Veris® sensor with maps produced from conventional grid-based sampling with one sample per hectare. Results showed that the extra detail provided by the Veris® maps generated an average $14.32/acre advantage in improved lime accuracy, based on a $30/ton applied cost of lime, not accounting for any yield benefit. The authors noted that the typical cost of a Veris® MSP3 map at this time was $15–20/acre which means the lime accuracy improvement alone covered most of the initial map cost.

7.4 Precision Nutrition Based on Measurement and Mapping of Crop Properties

Introduction

Crop sensing and yield mapping can help assess crop variability. Crop sensing is simply the process of using sensors to collect information about a growing crop. This section focusses on the most common commercially used form of crop sensing—crop reflectance measurements, where optical sensors are used to measure the amount of light radiation reflected from the crop at different wavelengths. A growing crop will absorb, transmit and reflect incoming radiation. The proportion of radiation absorbed, transmitted or reflected will vary across the electromagnetic spectrum (at different wavelengths) depending on crop characteristics. Most crop reflectance sensors measure reflectance in the visible (about 400–700 nm) and near infrared portion (about 700–2500 nm) of the electromagnetic spectrum. The variation in reflectance at different wavelengths is called the crops spectral signature. A healthy well-developed crop will have a different spectral signature to a less well developed or unhealthy crop. In this way, crop reflectance measurements can be used to provide information on crop growth and vigour.

Differences in reflectance at specific wavelengths can be expressed as a vegetation index—these can be calculated in many ways, but the most well-known is the Normalised Difference Vegetation Index (NDVI). Since reflectance from the crop is determined by the size and vigour of its canopy, vegetation indices have been shown to correlate well with crop characteristics such as above ground biomass, crop vigour, and chlorophyll content.

Crop sensing can provide valuable information on the performance of the crop across the entire planted area (spatial) and throughout the growing season (temporal) and can be used as the basis for variable rate nitrogen management, which has the potential to better target nitrogen fertiliser applications and improve crop nitrogen use efficiency.

Types of Crop Sensors

Optical crop sensors can be categorised according to the reflectance measurements recorded, the light source used, and the platform that the sensor is mounted on. The type of crop sensor affects how it can potentially be used to inform precision nitrogen management. The main types of crop sensors are detailed below.

Sensor Type According to Reflectance Measurements Made

Sensors can be categorised as RGB (Red, Green, Blue), multispectral, and hyperspectral based on the number of spectral bands they capture.

RGB sensors take an image in the visible (Red, Green, Blue) light spectrum, which recreates almost exactly what our eyes see. RGB images can be analysed at the pixel scale to identify objects (e.g. crop counting), to calculate the relative surface of an object of a given colour (e.g. % crop green colour) or to calculate a limited number of vegetation indices which use reflectance from the RGB wavebands. The main limitation of these sensors is that they do not provide crop reflectance from the near infrared waveband, which is needed to calculate some of the most useful vegetation indices, including most vegetation indices used for variable rate nitrogen management.

Multispectral sensors measure light from several wavebands simultaneously. By definition, a RGB camera is a form of multispectral camera. However, the term 'multispectral sensor' is normally used for a sensor that measures from both the visible and near infrared wavebands. Inclusion of the near infrared waveband allows calculation of vegetation indices that provide a better representation of crop growth and vigour. Multispectral cameras generally measure between 4 and 12 wavebands; inclusion of the green, red, red-edge and near infrared wavebands allows calculation of many vegetation indices related to crop biomass, vigour, and chlorophyll content. Most commercial variable rate nitrogen management services use multispectral sensors.

Hyperspectral sensors measure light in narrower (typically 5–10 nm) and more numerous bands than multispectral cameras. Hyperspectral sensors provide crop reflectance data across a continuous range of wavebands and therefore have a greater potential to detect differences in crop growth than multispectral sensors. However, hyperspectral cameras are more expensive and generate very large quantities of data to process and are therefore more commonly used in research rather than for informing farm management.

Passive and Active Sensors

Sensors can be classified as passive or active sensors depending on the light source used. Passive sensors use the sun as their light source and measure the amount of reflected sunlight. Because passive sensors rely on sunlight, reflectance measurements are affected by the angle of the sun and cloud cover, and it is generally recommended to take measurements close to solar noon when the sun is highest in the sky and when cloud cover is minimal. Some passive sensors try to minimise the impact of varying light conditions by also measuring incident light above the crop canopy to enable correction for varying light levels and/or frequent calibration of the sensor using a white reference panel.

Active sensors emit their own light source and measure the amount of light reflected to the sensor. Active sensors use a modified light source, which allows them to distinguish the reflectance from the background sunlight. This means that measurements are independent of changing background light levels and active sensors can be used at any time of day and without disturbance by passing cloud cover.

Sensor Platform

Sensors can be mounted on different platforms to collect the data. Measuring using satellite or drone mounted sensors is called remote sensing, whereas measuring using ground based hand-held, or tractor-mounted sensors is often referred to as proximal sensing as the sensor is closer to the object.

Handheld sensors provide a relatively quick and easy way of measuring reflectance from a limited number of points within a field. Some handheld sensors include a display, which shows crop reflectance measured in real time. Handheld sensors can be useful for a limited number of 'spot checks' within a field. However, it is not possible to provide crop reflectance information from larger areas using handheld sensors and therefore they cannot be used as a basis for prescription maps for variable rate nitrogen application.

Fig. 7.1 Fritzmeier ISARIA tractor-mounted crop sensor measuring crop reflectance from a cabbage crop

Commercially available tractor-mounted sensors are typically mounted on the cab or at the front of the vehicle and can be used to vary nitrogen applications 'on-the-go', for example the Yara N Sensor or Fritzmeier ISARIA sensor (Fig. 7.1). Tractor-mounted sensors typically record crop reflectance data from an area either side of the field tramlines and will not cover the whole field area. The resolution of the data collected will depend on the width of the tramlines used by the tractor. The software available with most commercially available tractor-mounted sensors then uses data interpolation methods to provide a complete crop reflectance map of the field.

Unmanned aerial vehicles (UAVs), also commonly known as drones, can be used to collect high resolution imagery from whole field areas. The flight path of the UAV is controlled either by pre-programmed onboard computer or by the remote control of a pilot on the ground. The UAV captures multiple overlapping images of the areas of interest, which are then combined or 'stitched' together into a single orthomosaic map for each waveband measured using photogrammetry software. The spatial resolution or Ground Sampling Distance (GSD) is the distance between two consecutive pixel centres measured on the ground. The GSD depends on the sensor model and flight height but is typically around 4 cm GSD at 60 m flight height and 8 cm GSD at 120 m flight height. It will typically take around 30 min to collect data from a 10-ha field using a drone flown at 120 m with 75% image overlap. Whilst the high spatial resolution available from UAV sensors is an advantage of this platform, UAV surveys can produce large amounts of data that can be time consuming to download and process. Multispectral (5-band) imagery from a UAV survey at 120 m flight height for a 10-ha field will typically be 150–200 MB when stitched together into a single image for each waveband. Most UAV mounted sensors are multispectral; hyperspectral sensors are available, but they significantly increase the volume of data to process. The main advantage of UAV data is the high data resolution, however this high level of data resolution is not needed for variable rate nitrogen applications as it is not possible to vary the nitrogen rate with this level of accuracy using commercial fertiliser spreaders, and therefore it is generally not economic to use UAV data for variable rate nitrogen application, where other cheaper options are available.

Satellite imagery is collected remotely and is relatively cheap and easy to access compared to other methods of crop sensing. The main limitations of satellite sensors are lower image resolution, frequency of measurements and the influence of cloud cover. Most satellite sensors are passive sensors which cannot penetrate cloud cover, which means there will be no usable data for areas of the image covered by cloud. Free satellite imagery is available at medium resolution (10–30 m) from ESA's Sentinel 2 and NASA's Landsat 8. Higher resolution data (5 m–50 cm) is commercially available for purchase. As the resolution and frequency of data acquisition has increased, commercial use of satellite data for variable rate nitrogen application has increased and there are now many precision farming companies offering this as a commercial service. Greater frequency of data acquisition helps address the issue of cloud cover, and some companies are starting to include radar imagery, which is unaffected by cloud cover.

Vegetation Indices

A vegetation index is a numerical value calculated as a mathematical combination of reflectance values measured at two or more wavelengths, which can be related to crop characteristics such as above ground biomass, crop vigour or chlorophyll content. Vegetation indices simplify the comparison of canopy spectral signatures and cheaper instruments have been developed that allow more rapid assessment of canopy reflectance at just two or three individual wavelengths (Kindred et al. 2016). Many vegetation indices have been developed and are described in the scientific literature. Different vegetation indices have been developed to be sensitive to specific plant characteristics such as biomass, nitrogen content or crop moisture status. The Index Database website (https://www.indexdatabase.de/) provides one of the most comprehensive databases of remote sensing indices.

The Normalised Difference Vegetation Index (NDVI) (Rouse et al. 1974) is perhaps the most used and well-known vegetation index. It is calculated as a ratio between reflectance in the red and near infrared wavebands. The NDVI formula will provide a value between − 1.0 and + 1.0; water has a negative value due to its strong absorption of near infrared radiation; soil surfaces typically have a value from 0.1 to 0.2 and cropped land from 0.2 to 1.0, with higher values for thicker crop covers. NDVI provides a good correlation with vegetation cover until the crop canopy closes and the Leaf Area Index (LAI) increases above about 3.0. Above this point, the response of NDVI to increasing vegetation flattens out or 'saturates'.

This saturation effect is well known, and other Vegetation Indices have been developed to try and provide a better measure of crop biomass at higher leaf area indices.

Several vegetation indices have been developed to be sensitive to leaf chlorophyll status, including Modified Chlorophyll Absorption in Reflectance Index 2 (MCARI2) (Haboudane et al. 2004), Meris Terrestrial Chlorophyll Index (MTCI) (Dash and Curran 2004), and Chlorophyll Index Green (CIgreen) (Gitelson et al. 2005). Leaf chlorophyll content provides an indicator of photosynthetic activity and therefore potential yield as well as a measure of the crop nitrogen status.

Variable Rate Nitrogen Management

Canopy sensing can be used as a basis for variable-rate nitrogen applications. Thinner, paler crops are associated with a lower soil nitrogen supply and may warrant extra nitrogen application in relation to thicker and greener areas of the crop (Kindred et al. 2016). This is the basis of most commercial variable rate systems, i.e. increasing nitrogen applications where growth is less and reducing nitrogen applications where crops are lush (Berntsen et al. 2006; Zillmann et al. 2006).

Tractor-mounted sensors such as the Yara N sensor and Fritzmeier ISARIA sensor can be used to vary nitrogen application rates on-the-go, whilst satellite or drone imagery can be used to provide a prescription map for variable rate nitrogen application. It is important to note that variable rate nitrogen as a crop management strategy assumes that crop variability measured by the sensor is at least partly due to variability in nitrogen availability. Nitrogen is one of the key factors affecting crop growth and yield. However, it is not the only factor. It is important to know the cause of the spatial variation seen to have confidence that it can be rectified by varying nitrogen fertiliser application rate. Where crop variability is due to other factors, such as pest or disease pressure, soil compaction, or soil moisture content, then varying the nitrogen rate is unlikely to be of benefit.

Another important consideration is the amount by which the nitrogen rate should be varied. For certain crops, some methods of crop sensing can provide a nitrogen rate based on crop canopy status without any further grower input. For example, the Yara N sensor has an absolute calibration for oilseed rape and the Fritzmeier ISARIA sensor has an absolute calibration for wheat. However, in other cases, growers will need to perform a field calibration by taking an area of crop in the field and allocating a target nitrogen rate to that area, and will also need to specify minimum and maximum nitrogen rates for the rest of the field.

The principles of variable rate nitrogen management are applicable across a range of crops, and this technology may have the potential to improve nitrogen use efficiency in horticultural crops. Variable rate nitrogen is most likely to be beneficial in crops that have a high nitrogen requirement and where most of the nitrogen is applied to the growing crop (rather than at planting), for example longer season brassica crops. However, unlike for arable crops, there is little guidance available to horticultural growers on how to use variable rate nitrogen in these crop types, and specifically by how much to vary nitrogen rates.

Most of the research work to date on variable rate nitrogen has focussed on arable crops, most commonly wheat and oilseed rape. Colaço and Bramley (2018) reviewed 24 studies that investigated the use of crop sensing to improve nitrogen management in maize, rice, wheat and oilseed rape crops. Most of the studies reviewed reported nitrogen fertiliser savings of between 5% and 45% with little effect on yield, and the average increase in profit reported was around $30/ha, although negative returns were also common. The authors noted that a lack of consistent evidence of economic benefit was limiting adoption by growers, and that most of the current sensor-based nitrogen strategies focus on delivering a simple solution to a complex problem.

The complexity of precision nitrogen management was further highlighted by Kindred et al. (2016) in the 'Auto-N' project. Novel chessboard nitrogen response experiments in winter wheat were set up on six commercial fields in the UK between harvest years 2010 and 2012 to quantify spatial variation in nitrogen requirement, to explain it in terms of crop nitrogen demand, soil nitrogen supply and fertiliser nitrogen recovery. Within-field variation in optimum nitrogen rates exceeded 100 kg/ha at all sites; spatial variation in optimal yield was greater than 2 t/ha at all sites, and variation in soil nitrogen supply was generally greater than 50 kg/ha. Some of the spatial variation in optimum nitrogen was explained in terms of soil nitrogen supply and crop nitrogen demand. However, the tendency for positive correlations between soil nitrogen supply and optimum yield was striking and hindered complete explanation of spatial variation in optimum nitrogen: i.e. high yielding areas tended to have greater soil nitrogen supply, so the increased requirement from higher crop nitrogen demand was counteracted by the reduced requirement from higher soil nitrogen supply. Further validation trials set up in 11 fields in 2013 and 2014 demonstrated the feasibility of automating judgements of nitrogen requirements within fields using precision information. However, the variability in crop nitrogen demand, soil nitrogen supply and fertiliser nitrogen recovery, and crucially the interactions between them, meant that the use of such systems would not guarantee increased accuracy or precision of nitrogen use. The evidence from this project suggests that variable rate nitrogen management can give only modest returns, even with a system making perfect predictions, if the field is already receiving the right average nitrogen rate. Kindred et al. (2016) concluded that the most important decisions concern nitrogen use for whole farms, then for whole fields, then for areas within fields. Precision technologies can help with all of these, especially through comparisons of crops between and within farms. However, the most effective aspect of precision farming technologies is probably the empowerment of growers to test retrospectively the effects of their nitrogen decisions (or indeed any decisions) on-farm. Given the variation in and unpredictability of nitrogen requirements between fields and between farms the only way growers can know for sure whether their chosen nitrogen rates were right is to test yield effects of different nitrogen rates—this is relatively easy now, by simply applying (say) 60 kg/ha more and 60 kg/ha less to adjacent tramlines.

Case Study: Variable Rate Nitrogen in Savoy Cabbage

A field demonstration experiment was set up to investigate the potential for canopy sensing for variable rate nitrogen applications to a Savoy cabbage crop in 2016 in Nottinghamshire, England. The demonstration included nitrogen response experiments and tramline comparisons of uniform and variable rate nitrogen application.

Nitrogen response experiments (seven nitrogen application rates between 0 and 360 kg N/ha) were replicated in three different areas of the field to see if there was any evidence of within-field variation in optimum nitrogen rate. In addition, the farm standard uniform nitrogen application rate (240 kg N/ha) was compared to variable nitrogen application (+/− 40 kg N/ha from the farm standard) in a tramline comparison.

Crop canopy measurements from the nitrogen response plots showed a strong relationship between NDVI and total biomass, and between NDVI and crop nitrogen uptake. The nitrogen response plots also showed a significant yield response to nitrogen fertiliser; fresh weight marketable yield increased from 10 t/ha on the zero-nitrogen treatment to a maximum of 44 t/ha at the highest nitrogen rate of 360 kg N/ha.

The large yield response to nitrogen fertiliser and the strong relationship between crop canopy measurements (NDVI) and above ground crop biomass and nitrogen uptake indicate that canopy sensing can be used as a basis to vary nitrogen applications, and where nitrogen availability varies across the field, we could potentially see 'evening-up' of the crop and/or an overall yield increase. However, statistical analysis of nitrogen response data from the three experiments showed that, in this case, nitrogen response was similar across the field (Fig. 7.2).

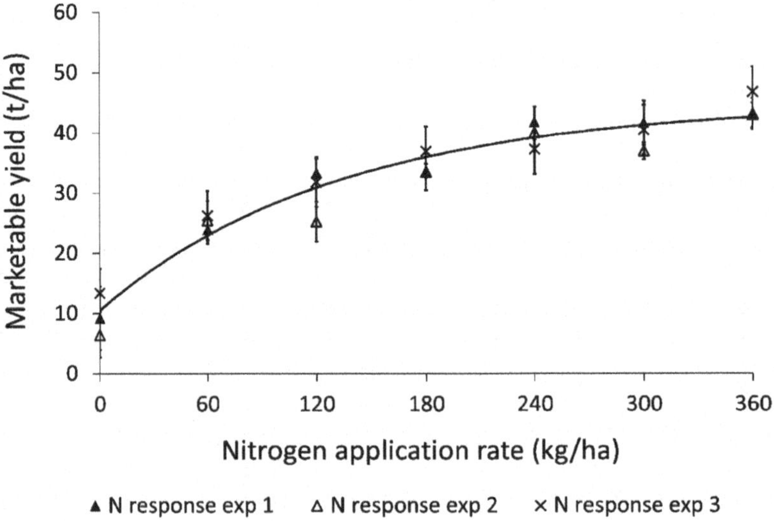

Fig. 7.2 Marketable yield response of Savoy cabbage to nitrogen fertiliser from three nitrogen response experiments in the same field

Comparison of marketable head weights and total marketable yields from the uniform and variable rate nitrogen tramline comparisons did not provide any evidence that varying the nitrogen rate increased total marketable yield or produced a more consistent sized crop in this Savoy cabbage crop.

Variable-rate nitrogen management will be of benefit only if nitrogen is the main factor causing variability in the crop canopy. At this site, the nitrogen-response experiments showed that the crop response to nitrogen was similar across the length of the field and the tramline comparisons did not show a yield benefit from variable-rate nitrogen management. It is likely that variability in the crop canopy here was due to other soil or crop factors i.e., not only due to nitrogen.

Uptake and Outlook

Crop sensors can provide valuable information on crop variability and performance. This technology is now readily available to growers, most commonly as either tractor-mounted sensors or via satellite imagery. The increase in spatial and temporal resolution of satellite data has made this data readily available to growers. Many precision farming companies offer variable rate nitrogen services using either tractor-mounted sensor or satellite data. However, uptake of and confidence in variable rate nitrogen as a management strategy has been limited by lack of clear and consistent evidence of yield and economic benefits. This may in part be because nitrogen availability is only one of many factors potentially contributing to crop variability.

A recent telephone survey of UK potato and horticultural growers (Sagoo et al. 2024) found that 10% (3 of 33 growers surveyed) had used variable rate nitrogen on their potato or horticultural crops; one on lettuce, one on sweetcorn, and one on brassica vegetables (cauliflower, broccoli, and cabbage). A third (11 of 33 growers) had used variable rate nitrogen on other crops on the farm, most commonly cereals. Just under half of growers (15) said they would know how to use variable rate nitrogen application as a management strategy for the crop in question. Two growers noted that whilst they understood how variable rate nitrogen management worked, they would be unsure about how much to vary the nitrogen rate. Only nine growers said they thought variable rate nitrogen application would be effective at increasing nitrogen use efficiency for the crop in question, 17 growers said they did not think it would increase nitrogen use efficiency, and 10 growers said they were unsure if it would increase nitrogen use efficiency.

Both Kindred et al. (2016) and Colaço and Bramley (2018) highlighted the complexity of precision nitrogen management, with Colaço and Bramley concluding that most of the current sensor-based N strategies focus on delivering a simple solution to a complex problem. However, this does not mean that the technology does not have the potential to improve nitrogen use, and Colaço and Bramley recommended that future developments should be based on a multi-sensor and multi-layered information approach implemented in a single algorithm, ideally delivered by means of an automatic sensing and application platform. Kindred et al. (2016) also made an important point about the importance of focussing on the areas where the most gains in efficiency could be made. From an economic perspective, it appears most important to achieve an accurate average nitrogen rate at the largest scale before accounting for variation at a smaller scale. Small errors have small costs, but large errors have disproportionately large costs. This is a particularly important point for horticultural growers, as nitrogen recommendations for horticultural crops (where they exist) are generally based on far more limited data sets of nitrogen response experiments than for arable crops. Therefore, growers may be best advised to focus their efforts on improving the accuracy of their average field nitrogen rate, rather than on varying nitrogen rates within fields.

7.5 Concluding Remarks

Soil variability is one of the key factors determining differences in crop yields within and between fields. Precision farming tools such as soil mapping and canopy sensing can provide growers with valuable information about the variability of their soils and crops. Where growers have identified variability in their soil or crop, they should first seek to understand what factors are important in causing this variability before they try to manage it.

Precision (variable rate) application of fertiliser and lime is one of the most common applications in precision horticulture. Potential benefits include increased crop yields and quality from applying the correct amount of fertiliser or lime to areas that would otherwise have been under fertilised, savings in fertiliser or lime from not overapplying to areas with high soil nutrient/pH levels, and associated potential environmental benefits.

One of the main barriers for further uptake of variable rate fertiliser applications is the lack of evidence to demonstrate that this approach can increase yields and profitability. Many horticultural fields have high background levels of soil fertility from high historic applications of fertiliser and therefore yield increases from variable rate application may not be expected. In these cases, the main benefit will be savings from reduced fertiliser application rates. However, small savings in fertiliser are generally less important to horticultural growers than to arable growers where the margins are generally smaller. There is also sometimes a reticence from horticultural growers to reduce fertiliser application rates, even to parts of a field, due to concern about potential yield loss from any fertiliser reduction. This may be linked to a lack of confidence in the fertiliser recommendations for these crops, which are often based on a limited number of response experiments. Whilst small savings from reduced fertiliser application rates may not be sufficient motivation for some horticultural growers to adopt variable rate technology, the increased focus on environmental impacts and legislation and other incentives in many countries to reduce nutrient losses to the environment may become an important driver for adoption in future years.

The technology for soil mapping and variable rate application of lime, and phosphorus and potassium fertiliser are well established and commercially available. Future work should focus on developing technologies that will allow more rapid, in-field, on-the-go sensing of soil phosphorus and potassium, provide higher resolution soil nutrient maps, and thus in turn improve the accuracy of variable rate phosphorus and potassium application.

Crop sensing for variable rate nitrogen management is more complex. The technology is well developed and commercially available to growers, however interpretation of the data is more difficult. Studies cited in this chapter have highlighted that a simple translation of a vegetation index map to a prescription variable rate nitrogen map may have limited yield or economic benefit if it does not correctly address the factors causing the measured crop variability. More work is required to understand how to interpret this data and how to best combine it with other data to better target nitrogen management.

Competing Interests The authors have no conflicts of interest to declare that are relevant to the content of this chapter.

References

Adamchuck VI, Lund ED, Sethuramasamyraja B, Morgan MT, Dobermann A, Marx DB (2005) Direct measurement of soil chemical properties on-the-go using ion-selective electrodes. Comput Electron Agric 48:272–294

Adamchuk VI, Hummel JW, Morgan MT, Upadhyaya SK (2004) On-the-go soil sensors for precision agriculture. Comput Electron Agric 44:71–91

AHDB (2023) Nutrient Management Guide (RB209). Section 1 Principles of nutrient management and fertiliser use. Agriculture and Horticulture Development Board. https://ahdb.org.uk/knowledge-library/rb209-section-1-principles-of-nutrient-management-and-fertiliser-use

Berntsen J, Thomsen A, Schelde K, Hansen OM, Knudsen L, Broge N, Hougaard H, Hørfarter R (2006) Algorithms for sensor-based redistribution of nitrogen fertilizer in winter wheat. Precis Agric 7:65–83

Caesar L, Sakschewski B, Seaby Andersen L, Beringer T, Braun J, Dennis D, Gerten D, Heilemann A, Kaiser J, Kitzmann NH, Loriani S, Lucht W, Ludescher J, Martin M, Mathesius S, Paolucci A, te Wierik S, Rockström J (2024) Planetary health check. https://www.planetaryhealthcheck.org/boundary-pages/modification-of-biogeochemical-flows. Accessed 31 Mar 2025

Chang CW, Laird DA, Mausbach MJ, Hurburgh CR (2001) Near-infrared reflectance spectroscopy–principal components regression analyses of soil properties. Soil Sci Soc Am J 65:480–490

Colaço AF, Bramley RGV (2018) Do crop sensors promote improved nitrogen management in grain crops? Field Crop Res 218:126–140

Dash J, Curran PJ (2004) MTCI: the MERIS terrestrial chlorophyll index. Int J Remote Sens 25(23): 5403–5413. https://doi.org/10.1080/0143116042000274015

FAO (2024) Inorganic fertilizers 2002–2022. https://openknowledge.fao.org/handle/20.500.14283/cd1485en

Gitelson AA, Viña A, Ciganda V, Rundquist DC, Arkebauer TJ (2005) Remote estimation of canopy chlorophyll content in crops. Geophys Res Lett 32(8). https://doi.org/10.1029/2005GL022688

Haboudane D, Miller JR, Pattey E, Zarco-Tejada PJ, Strachan IB (2004) Hyperspectral vegetation indices and novel algorithms for predicting green LAI of crop canopies: modelling and validation in the content of precision agriculture. Remote Sens Environ 90:337–352

He Y, Huang M, García A, Hernández A, Song H (2007) Prediction of soil macronutrients content using near-infrared spectroscopy. Comput Electron Agric 58(2):144–153

Heege HJ (2013) Site-specific fertilizing. In: Heege HJ (ed) Precision in crop farming. Site specific concepts and sensing methods: applications and results. Springer

Kindred DR, Storer K, Hatley D, Ginsburg D, Catalayud A, Milne A, Marchant B, Miller P, Sylvester-Bradley R (2016) Automating nitrogen fertiliser management for cereals (Auto-N) LINK Research Project LK09134: reducing GHG emissions, nitrate pollution and lost productivity by fully automating N fertiliser management. AHDB cereals and oilseeds final project report RD-2008-3530

Kulczycki G, Grocholski P (2013) The effect of long-term phosphorus and potassium precision fertilization. In: Precision agriculture, vol 13. Wageningen, Wageningen Academic, pp 395–401

Lee KS, Lee DH, Sudduth KA, Chung SO, Kitchen NR, Drummond ST (2009) Wavelength identification and diffuse reflectance estimation for surface and profile soil properties. Trans ASABE 52(3):683–695. https://doi.org/10.13031/2013.27385

Leip A, Britz W, Weiss F, de Vries W (2011) Farm, land, and soil nitrogen budgets for agriculture in Europe calculated with CAPRI. Environ Pollut 159(11):3242–3252. https://doi.org/10.1016/j.envpol.2011.01.040

Limpus S, O'Halloran J, Layden I (2016) Adoption of variable rate technology in Queensland's intensive vegetable production systems. Final report for GMX-INNOV-312. State of Queensland

Longchamps L, Tisseyre B, Taylor J, Sagoo L, Momin A, Fountas S, Ampatzidis Y, Schueller JK, Khosla R (2022) Yield sensing technologies for perennial and annual horticultural crops: a review. Precis Agric 23:2407–2448. https://doi.org/10.1007/s11119-022-09906-2

Lund ED, Collings KL, Drummond PE, Christy CD (2004) Managing pH variability with on-the-go pH mapping. In: Proceedings of 7th international conference on precision agriculture. Minneapolis

Maleki MR, Mouazena AM, De Ketelaerea B, Ramona H, De Baerdemaekera J (2008) On-the-go variable-rate phosphorus fertilisation based on a visible and near-infrared soil sensor. Biosyst Eng 99:35–46

Mallarino A (1998) Using precision agriculture to improve soil fertility management and on-farm research. https://crops.extension.iastate.edu/encyclopedia/using-precision-agriculture-improve-soil-fertility-management-and-farm-research. Accessed 23 Jan 2025

McBride MB (2022) Estimating soil chemical properties by diffuse reflectance spectroscopy: promise versus reality. Eur J Soil Sci 73(1):13192

Rouse JW, Haas RH, Schell JA, Deering DW (1974) Monitoring vegetation systems in the Great Plains with ERTS. 1974. In: Proceedings, 3rd Earth Resource Technology Satellite (ERTS) symposium, vol. 1, pp 48–62

Sagoo E, Newell Price P, White C (2017) The application of precision farming technologies to drive sustainable intensification in horticulture cropping systems: a review of precision farming techniques for improved soil and nutrient management. AHDB. https://projectblue.blob.core.windows.net/media/Default/Research%20Papers/Horticulture/CP%20107c%20PF-Hort%20Technology%20review.pdf

Sagoo E, Barnes B, Dowers J (2024) Efficient use of nutrients on potato and horticultural crops in times of high fertiliser prices. Final report to Defra for project NM0106. https://sciencesearch.defra.gov.uk/ProjectDetails?ProjectId=21694. Accessed 23 Jan 2025

Shah S, Hookway S, Wilkinson S, Neish E, Hoskins C, Fletcher M (2017) The effect of variable rate application of phosphate, potassium and coated fertilisers on soil indices and crop yield. Asp Appl Biol 135:45–58

Syers JK, Johnston AE, Curtin D (2008) Efficiency of soil and fertilizer phosphorus use: reconciling changing concepts of soil phosphorus behaviour with agronomic information. In: FAO fertilizer and plant nutrition bulletin, vol 18. Food and Agriculture Organization of the United Nations (FAO), Rome

Zillmann E, Graeff S, Link J, Batchelor WD, Claupein W (2006) Assessment of cereal nitrogen requirements derived by optical on-the-go sensors on heterogeneous soils. Agron J 98(3): 682–690

Chapter 8
Robotics in Precision Horticulture

Mike Duke, Shen Hin Lim, Ben McGuiness, Chi Kit Au, and Roger Robson-Williams

Abstract Robotics are becoming increasingly commonplace in many industries but their application in horticultural production systems is relatively poorly developed. There is considerable potential for horticultural robotics to address labour shortages, reduce work-related injuries, maximise productivity and minimise waste. However, there are also considerable technical challenges to be overcome. This chapter takes a detailed look at harvesting in apples and kiwifruit, and pollination in kiwifruit, to explore some of the most intractable problems in automating horticultural production. A case study of harvesting a horticultural row crop is also provided.

Keywords Robotics · Precision horticulture · Robotic harvesting · Automation in horticulture

8.1 Introduction

The application of robotics in the manufacturing, healthcare and transport sectors is now well-established e.g. Frohm et al. (2008), Davenport and Kalakota (2019) and Daily et al. (2017). However, commercially viable agricultural and horticultural robots are still relatively limited in both prevalence and application despite the potential benefits in terms of alleviating labour shortages (Kirsten Lovelock and Leopold 2009; Singh et al. 2020) and replacing humans for certain repetitive, physically demanding and sometimes hazardous tasks.

The need for automation has never been greater. Maximising the productivity of food production and minimising food waste is a priority for governments globally in the face of population growth (UN Department of Economic and Social Affairs

---

M. Duke · S. H. Lim · B. McGuiness · C. K. Au (✉)
The University of Waikato, Hamilton, New Zealand
e-mail: chikit.au@waikato.ac.nz

R. Robson-Williams
New Zealand Institute of Plant & Food Research, Auckland Mail Centre, Auckland, New Zealand

P. Newell Price et al. (eds.), *Precision Agriculture in Field Horticulture*, Progress in Precision Agriculture, https://doi.org/10.1007/978-3-031-96534-0_8

2017), climate change impacts, biodiversity loss, and severe perturbations in global supply chains and international trade. Consequently, over recent years a large amount of research in this sector has been undertaken (Balafoutis et al. 2020; Bramley and Ouzman 2019; Shamshiri et al. 2018; Bogue 2016; De-An et al. 2011) and various market leading suppliers of agricultural equipment currently manufacture highly automated tractors for autonomous navigation and precision planting of broadacre row crops. However, these machines are limited in their utility and are far from being capable of automating the many complex tasks involved in annual or perennial field horticultural production.

Many horticultural tasks inherit, to some degree, complexity from the natural world and will often require dexterity, delicacy, and accuracy. Hand-eye coordination enables human workers to collectively, and subconsciously, adapt their individual actions in response to visual and tactile feedback. It is this closed-loop feedback that enables them to perform the tasks effectively and efficiently. In contrast, robots, which take an open-loop approach, struggle to raise the task efficiency and effectiveness. This is a fundamental barrier to practical, high-speed automation of horticultural tasks in fields, vineyards, and orchards.

To automate such tasks, significant advancements in hardware, perception methods, control, and mechanical design must be made. Horticultural tasks often require complex, highly dexterous manipulation of unstructured environments, coupled with high performance sensing systems capable of identifying, and localising target crops at human-competitive work rates. Going forward, horticultural tasks will also require autonomy. The emergent field of horticultural robotics, a subset of precision horticulture, aims to address these task requirements with 'intelligent' machines; taking sensory information from the environment and processing this information to produce informed operation.

Arguably, outdoor production of perennial crops presents one of the greatest challenges and opportunities for robotics in horticulture. Accordingly, this chapter sets out the current state of the art in automating tree and vine fruit crop production, with a particular emphasis on harvesting. Automating this task is a priority for many fruit crops and presents some particularly intractable problems including those related to spatial distribution, variability in fruit maturity and occlusion by foliage and other vegetative growth. By way of contrast, a case study of asparagus harvesting, an outdoor row crop, is also presented. Similar complexity exists for glasshouse production, although some of the environmental features may be more amenable to automation.

8.2 Robotic Systems for Horticultural Tasks

A robotic system for an orchard or vineyard typically consists of four main elements:

1. A movable platform for the robotic system transportation;
2. A system for target identification and localization;

Table 8.1 Comparison of key horticultural tasks in an orchard

Task	End-effector	Operation	Target	Keep/dispose
Weeding	Gripper	Grab/cut	Weed	Dispose
Pollination	Spray nozzle	Spray	Flower	Keep
Thinning	Gripper	Grab/cut	Bud, flower or fruit	Dispose
Harvesting	Gripper	Grab/detach	Fruit	Keep
Pruning	Gripper	Grab/cut	Branch	Dispose

3. A robot arm consisting of a manipulator and an end-effector (a tool such as a gripper or blade) for performing the operation;
4. Provision for retaining harvested fruit on the robotic platform or transferring it elsewhere.

Many horticultural tasks, such as weeding, pollination, thinning (of buds, flowers or fruitlets), harvesting, and pruning, are typically repetitive and specific tasks. A task usually starts when the movable platform moves to a specific position in the orchard. The targets are then identified and localized so that the manipulator sends the end-effector to a target location sequentially to perform the operation until all the target positions are reached. The movable platform moves to the next location and the process repeats until the task is completed. In most situations, each horticultural task has its tailor-made end-effector. Table 8.1 lists the major end-effectors for some tasks and their operations. Of course, the target for each task is also different. Some of the targets are kept after the task, such as the fruit after harvesting, while the other targets are disposed of, such as the fruitlets after thinning.

Figures 8.1, 8.2, and 8.3 show three robotic systems for the horticultural tasks of harvesting and pollination. Figure 8.1 shows an apple harvesting robotic system for a planar cordon apple orchard with a close-up of the robot arm shown in the image in the top left of the figure (Hamish 2018). A kiwifruit harvesting robotic system (Williams et al. 2019) is shown in Fig. 8.2. Figure 8.3 presents a kiwifruit flower pollination robotic system (Barnett et al. 2017).

The following sections explore in some detail the technical considerations associated with the design of robotic arms (manipulator configuration) and attached tools (end-effectors). Also target localisation, scalability and work rate (task cycle time). The development of orchard systems suitable for automation is also covered briefly.

8.3 Manipulator Configuration

A workspace is the total volume enclosing the reachable locations of the end-effector as the manipulator executes all possible motions. The basic objective for designing a manipulator should be to ensure that its workspace is embedded in the canopy so that the end-effector can reach the target location (such as fruit or flower locations) for certain operations to complete a horticultural task (such as pollination or harvesting).

Fig. 8.1 An apple harvesting robot system with a close-up of the robot arm shown in the image in the top left of the figure

Fig. 8.2 A kiwifruit harvesting robotic system with four robot arms

Fig. 8.3 A kiwifruit pollination robotic system with two spray manifolds

From an operational standpoint, the robotic task process can be split into phases. A phase is the process such that the movable platform of the robotic system stops, targets are identified and localized, the robot arm attempts to perform the task for all reachable targets, then the movable platform advances to a new, incomplete region of canopy to repeat the task. If an orchard consists of N_{target} reachable target locations and requires N_{phase} phases to complete a task with all reachable targets, then the number of reachable targets per phase $\frac{N_{target}}{N_{phase}}$ is given by

$$\frac{\mathrm{N}_{\mathrm{target}}}{\mathrm{N}_{\mathrm{phase}}} = \rho_{\mathrm{v}} \mathrm{V}_{\mathrm{h}} \tag{8.1}$$

where ρ_V is the reachable target location distribution density;

V_h is the intersection volume between the canopy and the workspace of a manipulator.

Assuming that the distribution density $_V$ is homogeneous, a large intersection volume V_h yields a higher number of reachable target locations per phase.

For instance, the intersection of a tree canopy and workspace of an apple harvesting robotic system is depicted in Fig. 8.4. The manipulator of an apple harvesting robot has three rotary axes and one linear axis. The linear axis is the vertical gantry axis that provides the z-axis movement to reach the top of the canopy. In addition, the rotary axes of other links are all vertical. The manipulator of the apple harvesting robot yields a prismatic workspace, which generates a large intersection volume V_h with the canopy. The harvesting robot can be placed on both sides of the vertical linearized canopy so that most of the canopy is covered by the workspaces of the robot as it moves along the rows.

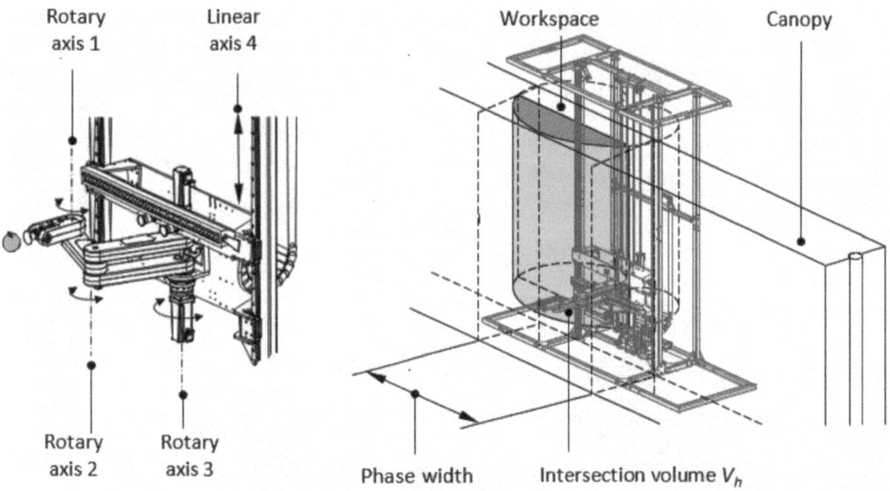

Fig. 8.4 A diagrammatic representation of a tree canopy and workspace of an apple harvesting robotic system. The image on the left shows a robot arm with a manipulator and end-effector. It consists of three rotary axes and one linear axis. The image on the right shows the intersection volume of the apple tree canopy and the workspace of the robot arm. The harvesting robotic system moves to the next phase after completing the harvesting in the current phase

In a pergola-style kiwifruit orchard, the canopy depicted in Fig. 8.5 is oriented horizontally above the harvesting system. The manipulator of the kiwifruit harvesting robotic system consists of four rotary axes. The base has two rotary axes. One of the axes is vertical and the other axis is horizontal, which aligns with the other axes within the manipulator. All the link lengths for both harvesting robots were developed based on the known dimensions of the fruit tree canopies and the space available for the robotic systems.

Despite the relatively large volume of workspace of a manipulator with rotary axes, the irregular toroidal workspace geometry means that the intersection volume between the canopy and workspace is limited. However, multiple robot arms can be employed to complete the harvesting task in a phase. This increases the width of the intersection volume by combining the workspaces. Furthermore, multiple intersection volumes can be combined by moving the robotic system across the phase. Consequently, the machine vision system must detect and localise fruit more frequently.

8.4 End-Effector Design

Each phase is broken down into a set of sub-phases. A sub-phase is the process where a robot arm moves to a new target location and performs an operation. There are several factors that influence sub-phases, including the motor/drive performance,

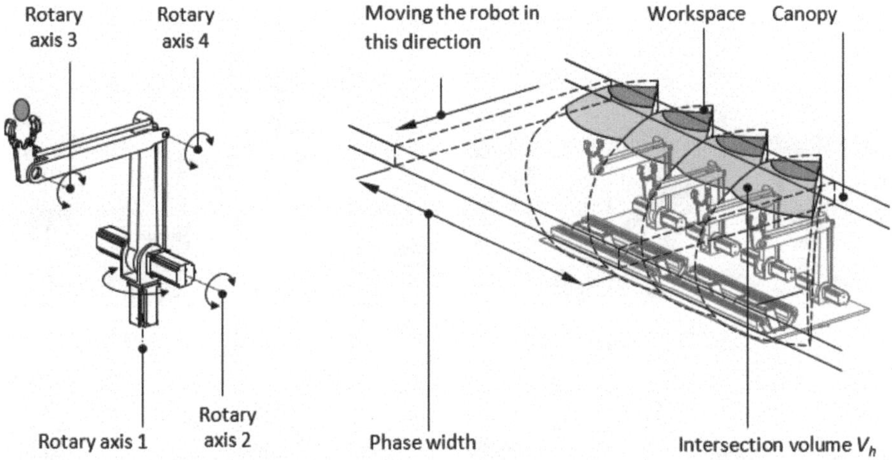

Fig. 8.5 A diagrammatic representation of a kiwifruit canopy and workspace of a kiwifruit harvesting robotic system. The image on the left shows a robot arm with a manipulator and end-effector. It consists of four rotary axes. The image on the right shows the intersection volume of the kiwifruit tree canopy and the combined workspace of four robot arms. Due to the smaller intersection volume, the harvesting robotic system moves along the phase advancement direction to complete the harvesting phase before it moves to the next phase

path planning, distance between target locations (scheduling), canopy disturbance and end-effector performance. The important considerations for end-effector design are dependent upon the types of horticultural tasks to be undertaken. For example, the considerations for a fruit harvesting gripper are variation in fruit size, variation in end-effector location, fruit clustering, and the imperative to minimise the risk of damage to the fruit.

Apples are harvested when they are ripe. To prevent bruising during harvesting, a suction cup is used to grab the fruit. It is important that a harvested apple should have the stem intact so that fruit loss can be prevented. An apple without a stem provides an entry point for micro-organisms, which can result in spoilage. The harvesting action consists of two rotational motions about the longitudinal and lateral axes. When a target fruit is grasped by the suction cup, a rolling motion along the longitudinal axis is performed first so that the fruit rotates about its centre for detachment. Rotating about the fruit centre generates a pulling force to detach the apple so that the stem remains in the apple. Then a pitch motion about the lateral axis is exercised to drop it into the corrugated pipe connecting to the collection bin as depicted in Fig. 8.6.

By contrast, kiwifruit are usually harvested while they are still hard. The end-effector uses a soft silicone air channel structure to conform to the fruit shape and maximise contact area. They can also deal with fruit shape, size, pose, and position variation through shearing of the supporting channels. The end-effector grabs a fruit and rotates it about the fruit-stem junction in a simultaneous motion, which is an effective and time-efficient method of detachment. The fruit-stem connection is much better at withstanding axial load than shear loading, hence if the kiwifruit

Fig. 8.6 The process of harvesting an apple using a robotic harvesting system

are pulled straight down along the stem-axis, the weakest point along the load path is often the stem-branch connection. Detachment at the stem-branch connection is unfavourable as fruit will be harvested with an attached stem. As the end-effector is unactuated and grip force relieved, the detached kiwifruit drops into a tube and onto a conveyor; this means that the robot-arm can continue to the next fruit without needing to place the fruit after detachment. The detachment process is shown in Fig. 8.7.

Figure 8.8 shows a robotic pollinator end-effector for kiwifruit flower pollination. It is a spray manifold with 40 solenoid-operated pollen delivery nozzles. The nozzles are arranged in a rectangular array in the manifold. It has a link with a linear sliding joint that is driven by a servomotor with an electric ball-screw cantilever actuator. As a result, an appropriate vertical distance between the flower and the spray manifold is maintained so that individual flowers within a kiwifruit canopy can be accurately sprayed.

8.5 Target Localization

Target localization in a robotic system is the foundation of robot intelligence. Target identification is necessary before the target is localized. This process can be performed by using machine learning techniques such as convolutional neural

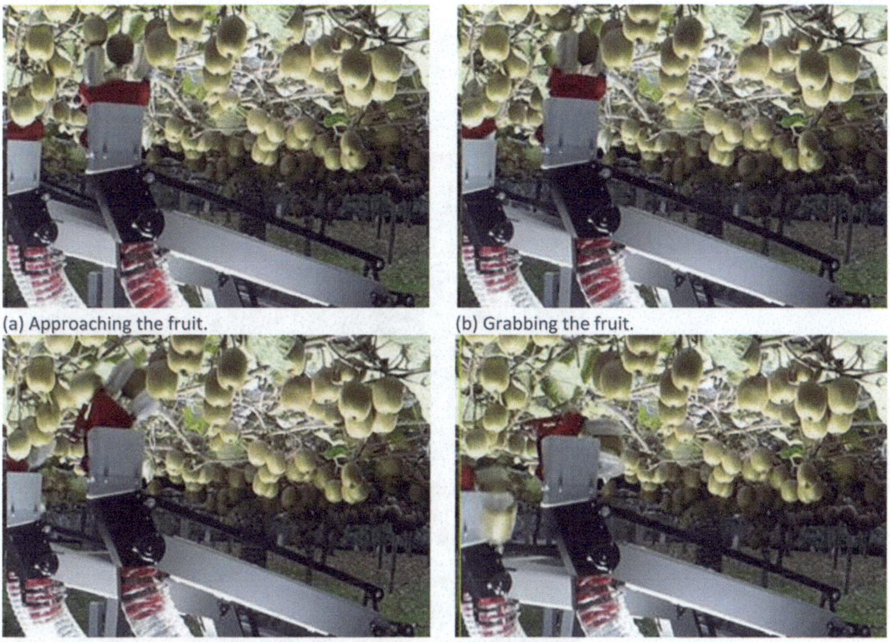

(a) Approaching the fruit. (b) Grabbing the fruit.

(c) Rotating the fruit about its stem junction (d) Dropping the fruit into a tube.

Fig. 8.7 The process of harvesting a kiwifruit using a robotic harvesting system

(a) Approaching the flowers. (b) Raising the manifold and spraying the flower.

Fig. 8.8 The process of pollinating kiwifruit flowers using a robotic system

network (CNN; Torres et al. 2020). This is a class of deep neural networks that utilise both the data and structure of their inputs to generate inference. Such networks are often used for high level image processing tasks, such as classification and fruit identification. The neural network takes a single undistorted red, green and blue (RGB) image of the canopy as an input, and outputs a fixed number of "bounding boxes", each with an associated target prediction and confidence value. Figure 8.9 shows an output from a CNN for apple identification.

The confidence value associated with each bounding box provides a measure of certainty about whether the target is present in the subsection of the image bounded

Fig. 8.9 Apples identified by a CNN. The identified apples are bounded by bounding boxes with confidence values

by the bounding box. In practice, a confidence threshold is generally enforced so that only bounding boxes with high confidence values are considered valid. A well-trained model will produce bounding boxes with high confidence values that completely enclose all instances of the corresponding target in the image. Figure 8.10 is another example of kiwifruit flower identification using a CNN.

Once the target is identified, sensing devices, such as a stereo vision system, are required to localize it. Stereo vision (Tang et al. 2020) is an adopted range imaging technique that utilises a camera pair to localize a fruit in a canopy from two known perspectives. Both harvesting and pollination systems shown in Figs. 8.1, 8.2 and 8.3 have been equipped with a stereo vision system for target localization. When the transformations between these perspectives are known, the relative position of the target in each image can be used to infer its distance from the camera using the epipolar geometry. Figure 8.11 outlines the processes for localizing the targets in a canopy for a horticultural task using a stereo vision system.

Multiple stereo vision systems can be used to widen the sensing coverage. The target locations computed by each system are then fed into a scheduler for operation scheduling. This includes merging the target locations, allocating task targets to each of the robot arms and determining the operation order. Furthermore, the robot arm collision avoidance (with other robot arms or branches, trellising and other

Fig. 8.10 Kiwifruit flowers identified by a CNN. The confidence values of the identified flowers are not shown in this image

Fig. 8.11 Information flow in localizing targets using a stereo vision system

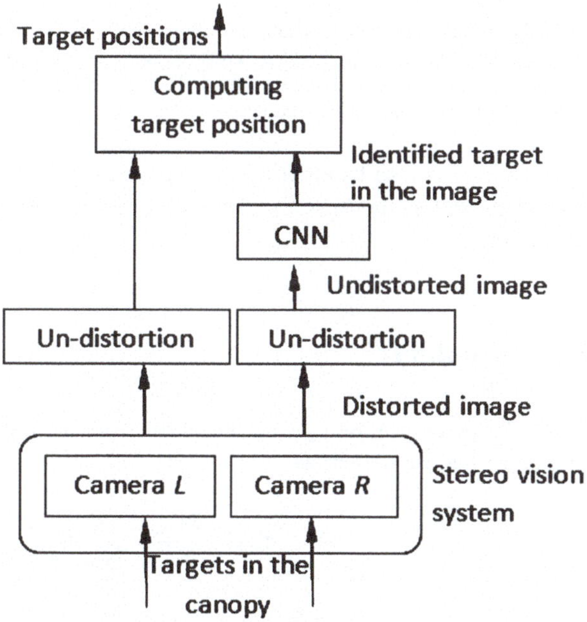

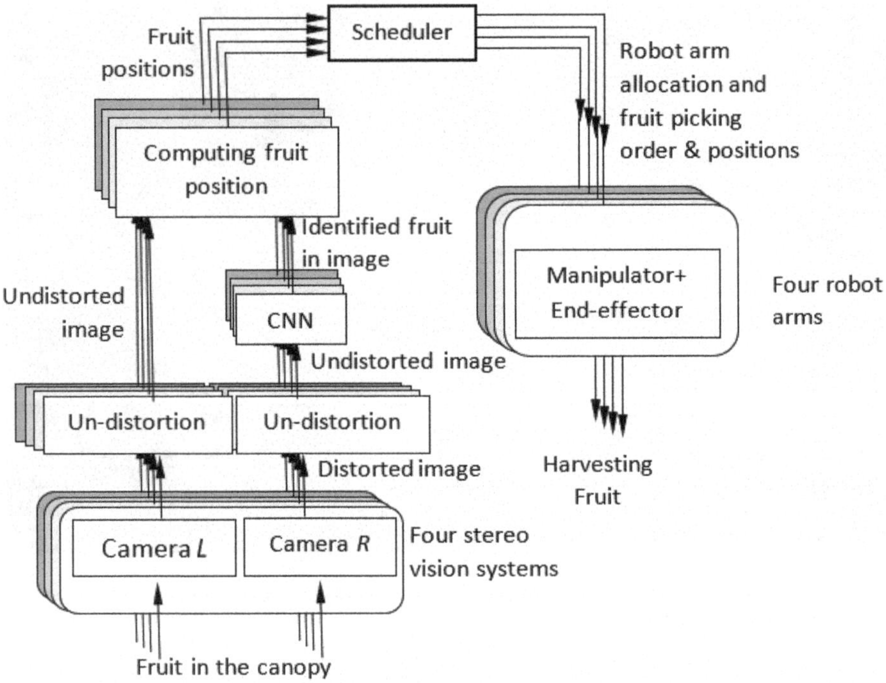

Fig. 8.12 Information flow in fruit localization using four stereo vision system and four robot arms. The fruit locations from each stereo vision system are merged together in the scheduler which allocates the harvesting to four robot arms. The allocation is based on minimizing the difference between work distribution W_D and unity

obstructions) is also handled by the scheduler. Figure 8.12 shows the information flow from four stereo vision systems to a centralized scheduler for fruit harvesting scheduling.

8.6 Scalability

The task operation time t_{task} for a multiple robotic system with N_{arms} manipulators in a phase can be written as

$$t_{task} = \frac{N_{target}}{N_{phase}} \frac{t_{sub-phase}}{N_{arms} W_D} \tag{8.2}$$

where $t_{sub-phase}$ is the sub-phase time that includes time for the end-effector to move from one target location to another target location and perform the operation. W_D is a unitless work distribution variable.

A variable, W_D, is proposed as a method for accounting for and evaluating a work distribution among the robotic-arms (Barnett et al. 2019). W_D is essentially introducing a component of operational down-time to a task phase. This unitless, normalized variable is equal to the ratio of average theoretical work distribution at parity to average work done by the busiest robot arm:

$$W_D = \frac{W_{parity}}{W_{busiest}} \tag{8.3}$$

The variable is defined as such because the busiest robotic arm is what delays the system. From Eq. (8.3), the maximum value of W_D is unity when $W_{parity} = W_{busiest}$. This implies that the workload allocated to each robot arm by the scheduler is evenly distributed within the multiple robot system. In fact, the work distribution W_D measures how much the busiest work distribution deviates from the uniform work distribution as the busiest robot arm usually delays the system from progressing to the next phase. Hence, multiple robot arms cooperate to complete the task. At the same time, each robot arm also competes with the others during the task execution. As a result, the productivity of a multiple robot arm system can be raised non-linearly.

8.7 Task Cycle Time

Task cycle time is important as it describes the rate at which the robot can work relative to manual labour. It is generally defined as the time taken to locate the target and perform the operation. By process analysis, the average task cycle time t_{CT} of a robotic system in a phase, with units of second per target, across a region of canopy in an orchard, can be expressed as:

$$t_{CT} = \frac{N_{phase}}{N_{target}}(t_{move} + t_{localize} + t_{task}) \tag{8.4}$$

where t_{move} is the time for the mobile platform to move from phase to phase; $t_{localize}$ is the time for the localization system to localize the target in a phase; t_{task} is the task operation time.

Subsisting Eq. (8.2) into Eq. (8.4) yields

$$t_{CT} = \frac{N_{phase}}{N_{target}}(t_{move} + t_{localize}) + \frac{t_{sub-phase}}{N_{arms} W_D} \tag{8.5}$$

Besides the hardware performance of a robotic system, such as the movable platform, motors and actuators of the manipulator and end-effector, this equation shows that the task cycle time of a robotic system across a region is affected by the arrangement of the canopy in an orchard and the configuration of the manipulator.

The canopy arrangement determines the number of harvesting phases N_{phase}. It also defines the intersection volume generated by the manipulator configuration to yield the number of task targets N_{target}. On the other hand, the number of arms N_{arms} and the work distribution W_D given by the scheduler account for the effects of scalability on the task cycle time.

8.8 Automation-Enabled Orchards

An important approach to improving the performance of robots in perennial horticulture is to design the growing system with automation in mind. In particular, simplifying the canopy of tree and vine crops through trellising and pruning to create a simpler, linear architecture.

For instance, the apple trees in a structured orchard are grown in straight and narrow rows so that the light interception is increased. Orchard automation machines can be positioned between the rows of the apple trees. Apple trees with twin leaders are custom grown in the nursery so that the leaders are trained down to near horizontal. These trees are then moved to the orchard. In planar cordon apple tree orchards, also known as the Future Orchard Planting Systems (FOPS; Plant & Food Research 2020), the apple trees are planted between two posts established in a row. The fruiting branches grow almost vertically within the row and a vertical linear canopy is established as shown in Fig. 8.13. As a result, a linearised canopy is oriented vertically and the areas for horticultural tasks, such as fruit harvesting, are

Fig. 8.13 An apple orchard with linearized canopies. The apple trees are planted between two posts arranged in rows

Fig. 8.14 A pergola-style kiwifruit orchard. The vines are grown next to the frame. The kiwifruit hangs free from the leaves and shoots below the pergola

arranged one by one along a row. Different systems that apply similar concepts to FOPS systems, such as two-dimensional horizontal or v-shaped trellis, are also suitable for orchard automation machines.

Kiwifruit are commonly grown in a structured orchard on the pergola frame for commercial cultivation as shown in Fig. 8.14. A pergola frame is essentially a rectangular array structure aimed at building a supporting network for the shoots and the fruit. The kiwifruit vines climb and are trained onto the pergola structure. The flowers and fruits then spread below the pergola, hanging free from the leaves and shoots. As a result, the canopy is linearized with the orientation approximately aligned with the ground. The pergola system provides access along the rows, between the pergolas' brackets, for mechanical equipment of various height. Therefore, the areas for horticultural tasks can be arranged in a two-dimensional array in two directions.

8.9 Concluding Remarks

Orchards are not the controlled and structured environments found in factories and hospitals where robots have been successfully deployed. Among other challenges associated with the outdoor environment, the target locations for a task are often not stationary. For instance, fruit may shift within a cluster when one fruit is plucked, or the flowers in a cluster may be disturbed by the impact of a spray application. These transmitted reaction forces between targets, combined with wind disturbance, occlusion and motion over uneven ground, create complex dynamic environments with constantly moving targets.

To overcome these challenges, the future development of horticultural robotic systems will need to concentrate on system integration among the components of the localization system, robot arm and movable platform so that the system performance and efficiency for various tasks can be improved. Without such integration, it is unlikely that horticultural robots will be able to match or exceed human operators regarding daily work rates.

In particular, it is likely to be necessary to move to a real-time localisation system. The operation of the robotic system at a specific zone follows a cycle in a phase as shown in Fig. 8.15a. The current robotic system is operated with an "offline" localization system. It just senses the task zone once at the beginning of a task cycle. Then the targets are identified and localized. The localized target positions are collated by a scheduler that assigns the target for each robot arm dynamically. When the robot arm controller receives a target location from the scheduler, the joint parameters are calculated and the manipulator drives the end-effector to the target location and performs the operation. Then the next target for the robot arm will be determined by the scheduler based on the initial outcomes of the localization system until all targets in the scheduler are assigned. Then the movable platform moves to another zone for the next phase. Disturbances caused by uneven ground, wind moving targets about and transmitted reaction forces among targets are ignored in current approaches to performing orchard tasks and result in poor performance, such as missing fruit while harvesting or inaccurate spraying during pollination application.

To account for this, one approach is to re-run the cycle as a "supplementary cycle" repeatedly to complete the operation at that zone. The flow charts of these processes are outlined in Fig. 8.15b. Unfortunately, this increases the cycle time and reduces the productivity of the robotic system.

In the future, it should be possible to seamlessly integrate the components within a horticultural robotic system by employing various sensing devices to collect and gather the environmental information continuously. This would allow the system to respond to, and compensate for, the perturbations that it will encounter in complex natural environments. As a result, a "real-time" localization system will be needed for the next development phase as depicted in Fig. 8.15c. With a real-time localisation system, instantaneous task information can be obtained, and the scheduler should be able to update the task operation order accordingly. Furthermore, the robot arm controller should be more accurate in sending the end-effector to a specific target location.

An additional consideration for the future development of horticultural robotics is that certain tasks, such as fruit harvesting, require good contact between the end-effector and the target fruit. For some fruit, such as apples, pears and tomatoes, the gripping mechanism can damage the fruit due to concentrated pressure points when

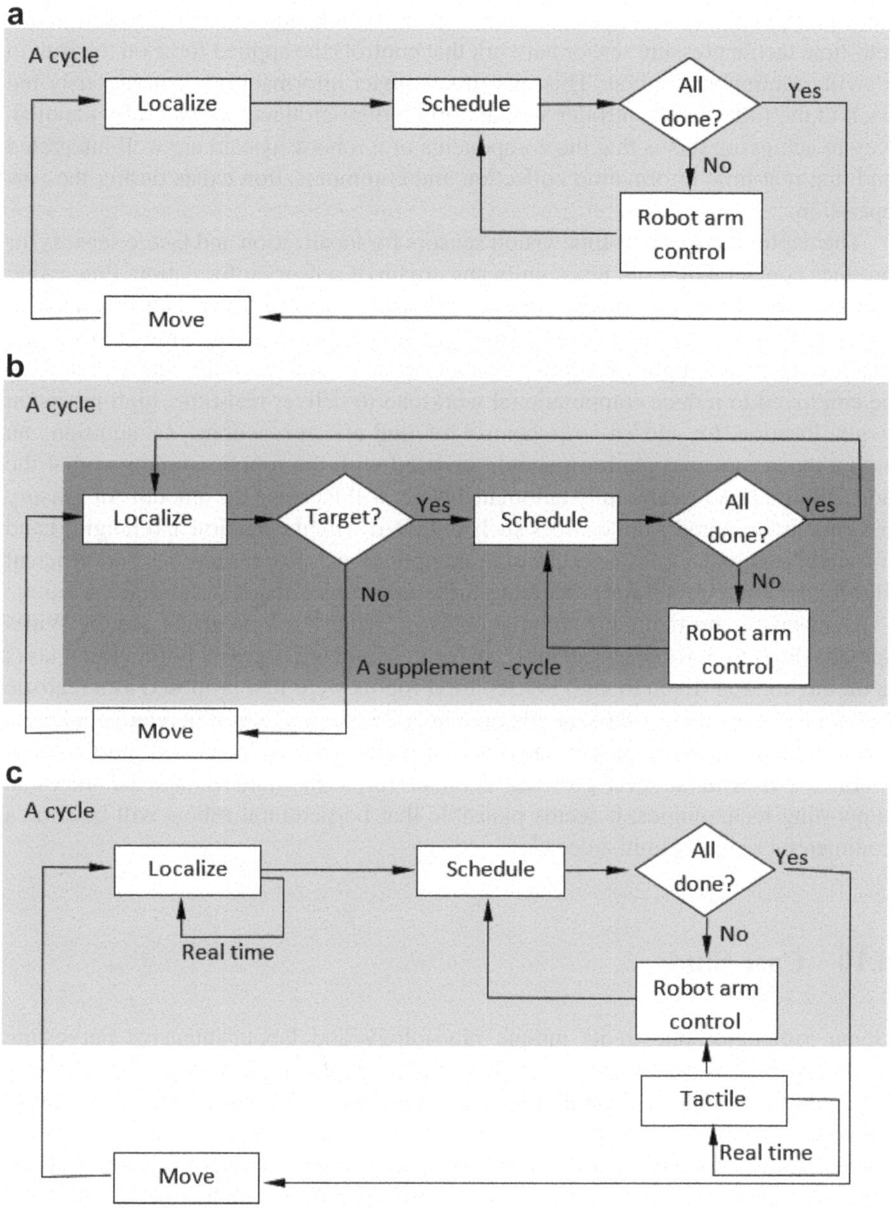

Fig. 8.15 (**a**) The cycle of a robotic system at a specific location: a cycle in a phase using an "offline" localization system. (**b**) The cycle of a robotic system at a specific location: a cycle in a phase using an "offline" localization system with a "supplement cycle". (**c**) The cycle of a robotic system at a specific location: a cycle in a phase using "real-time" localization system

force is applied. One solution is a soft end-effector embedded within a custom-made, real-time tactile pressure sensor network that controls the applied force on the fruit in a swift, continuous motion. This real-time contact information is continuously fed back to the robot arm controller so that horticultural products are carefully handled. Key to achieving this is that the components of a robotic system are well integrated and that real-time information collection and communication exists during the task operation.

The deployment of real-time vision sensors for localization and tactile sensors for produce contact will result in a significant stream of sensory information. Processing this must be done fast enough for the real-world orchard tasks with hundreds of thousands of targets per hectare. This will involve fusing data from videos and images, depth and high-speed pose-estimate sensors. Adaptive-filtering will need to be employed to reduce computational workload to deliver real-time, high-precision target location for closed-loop control of end-effector actions. In addition, an autonomous movable platform should be used with the robotic system so that the horticultural tasks can be fully automated. This will increase the amounts of sensory information required from devices such as LiDAR (light detection and ranging) and infrared rangefinders. As a result, high computational power facilities and efficient algorithms are required for processing these large amounts of sensory information.

Demand for horticultural automation is growing throughout the world. Widespread adoption of robotic systems, capable of achieving complex horticultural tasks is on the horizon. Even though horticultural robots were first proposed as a realistic possibility in the early 1980s, at this time in 2025 the field is still in relative infancy. It will take a multi-disciplinary approach to realise the full benefits of horticultural robotics but with an ever-growing demand for such systems, and advances in supporting technologies, it seems plausible that horticultural robots will become a commercial reality within a decade.

8.10 Case Study

Green asparagus, due to its unique physiology and labour-intensive harvesting requirements, has been a long-time crop of focus in this field. Despite the relatively large body of work surrounding the development of selective robotic asparagus harvesting solutions, no such machine has yet ascended to the level of being commercially competitive. A critical component of a prospective selective robotic asparagus harvester that is missing from contemporary machines is a robust vision system, capable of detecting, and localising, harvest-eligible spears in real-time from a commercial asparagus field.

To address this shortcoming a perception pipeline has been implemented (Peebles et al. 2020) using the time-of-flight camera as the vision sensor in a functional, prototype robotic harvester capable of harvesting green asparagus spears in a real-world commercial setting as shown in Fig. 8.16.

Fig. 8.16 A prototype asparagus harvesting robotic being towed by a tractor

A 100 ha asparagus farm typically consists of around 600 km of growing beds in multiple rows. This row length would require approximately 80 human workers to harvest the asparagus during peak season. Harvesting the asparagus using a robotic system can reduce labour cost. More importantly, it solves the issue of labour shortage.

The robotic harvester has a tractor-drawn carriage so that it straddles a single row of asparagus and travels down the direction of the row. An encoder is mounted at each wheel of the carriage for tracking the harvester. A time-of-flight camera is used as the vision sensor. A video and a point cloud are generated as it travels along the row. This information is used for identifying, localizing and categorising the oncoming harvestable spears by height. A manipulator with an end-effector is used to harvest the spears when their locations are received from the robot controller.

The prototype asparagus harvesting robot operates as follows:

1. The robot moves down a single asparagus row, drawn by a standard tractor at a constant speed.
2. As the robot moves down the row, the time-of-flight camera continuously streams data to a processing unit housing the vision system.
3. Wheel encoders continuously stream data to the processing unit as the robot moves down the row. The processing unit uses the information to keep track of the robot's position.
4. The processing unit utilises base predictions from the vision system, and positional data from the wheel encoders, to calculate when a target spear will be passing beneath the harvester. Commands are sent to the translation stage so that the end-effector is positioned to intercept the spear as the machine continues to move forward.

Fig. 8.17 A pointcloud of the scene is created from the data streamed from the time-of-flight camera as the vision system travels down the row of growing bed. The image in the top left of the figure shows the RGB video from the time-of-flight camera used for spear identification. The green boxes show the identified spears

5. The end-effector harvests the spear and transports it to storage.

Figure 8.17 shows the visualisation of the data streams from the time-of-flight camera. To achieve real-time performance, it is important that the required amount of data-processing is minimised.

The prototype harvester harvests 45.9% of all harvest eligible spears with a precision of 87.2% at its nominal ground speed of 0.3 m/s. The vision system detects 97% of all harvest-eligible spears with a precision of 74.5% at this speed. When operating at a ground speed of 0.7 m/s, the harvesting rate fell to 22% with 95% precision. The corresponding drop in the vision systems detection rate is 92.5% with 87.2% precision at a ground speed of 0.7 m/s. It is concluded that the vision system outperformed the available hardware when the ground speed is low. The localization system demonstrated a robustness to the unstructured nature of commercial asparagus fields, providing spear locations, which successfully facilitated the robotic harvest of green asparagus spears. As the ground speed increased, its performance dropped due to the limited computation power and the efficiency of the identification and localization algorithm.

Competing Interests The authors have no conflicts of interest to declare that are relevant to the content of this chapter.

References

Balafoutis AT, Van Evert FK, Fountas S (2020) Smart farming technology trends: economic and environmental effects, labor impact, and adoption readiness. Agronomy 10(5):743. https://doi.org/10.3390/agronomy10050743

Barnett J, Seabright M, Williams H, Nejati M, Scarfe A, Bell J, Jones M, Martinson P, Schare P, Duke M (2017) Robotic pollination—targeting kiwifruit flowers for commercial application. Presented at the PA17 International Tri-Conference for Precision Agriculture, Hamilton, New Zealand

Barnett J, Duke M, Au CK, Lim SH (2019) Work distribution of multiple Cartesian robot arms for kiwifruit harvesting. Comput Electron Agric 169. https://doi.org/10.1016/j.compag.2019.105202

Bogue R (2016) Robots poised to revolutionise agriculture. Ind Robot 43(5):450–456. https://doi.org/10.1108/IR-05-2016-0142

Bramley RGV, Ouzman J (2019) Farmer attitudes to the use of sensors and automation in fertilizer decision-making: nitrogen fertilization in the Australian grains sector. Precis Agric 20:157–175. https://doi.org/10.1007/s11119-018-9589-y

Daily M, Medasani S, Behringer R, Trivedi M (2017) Self-driving cars. Computer 50(12):18–23. https://doi.org/10.1109/MC.2017.4451204

Davenport T, Kalakota R (2019) The potential for artificial intelligence in healthcare. Future Hosp J 6(June):94–98. https://doi.org/10.7861/futurehosp.6-2-94

De-An Z, Jidong L, Wei J, Ying Z, Yu C (2011) Design and control of an apple harvesting robot. Biosyst Eng 110(2):112–122. https://doi.org/10.1016/j.biosystemseng.2011.07.005

Frohm J, Lindström V, Stahre J, Winroth M (2008) Levels of automation in manufacturing. Ergon Int J Ergon Hum Factors 30(3):181–207

Hamish C (2018) Development and evaluation of a robotic apple harvester to reduce fruit damage statistics. M. Eng. thesis. University of Waikato

Kirsten Lovelock K, Leopold T (2009) Labour force shortages in rural New Zealand: temporary migration and the recognised seasonal employer (RSE) work policy. N Z Popul Rev/N Z Demogr Soc 33(1):213–234

Peebles M, Barnett J, Duke M, Lim SH (2020) Robotic harvesting of asparagus using machine learning and time-of-flight imaging – overview of development and field trials. In: Conference: 2020 IEEE 16th international Conference on Automation Science and Engineering (CASE). https://doi.org/10.1109/CASE48305.2020.9217006

Plant & Food Research (2020). https://www.plantandfood.com/en-nz/article/robots-and-super-orchards

Shamshiri RR, Weltzien C, Cornelia, Hameed IA, Yule IJ, Grift TE, Balasundram SK, Pitonakova L, Ahmad D, Chowdhary G (2018) Research and development in agricultural robotics: a perspective of digital farming. Int J Agric Biol Eng 11(4):1–14. https://doi.org/10.25165/j.ijabe.20181104.4278

Singh B, Shirsath PB, Jat ML, McDonald AJ, Srivastava AK, Craufurd P, Rana DS, Singh AK, Chaudhari SK, Sharma PC, Singh R, Jat HS, Sidhu HS, Gerard B, Braun H (2020) Agricultural labor, COVID-19, and potential implications for food security and air quality in the breadbasket of India. Agric Syst 185. https://doi.org/10.1016/j.agsy.2020.102954

Tang Y, Chen M, Wang C, Luo L, Li J, Lian G, Zou X (2020) Recognition and localization methods for vision-based fruit picking robots: a review. Front Plant Sci 11:510. https://doi.org/10.3389/fpls.2020.00510

Torres JN, Mora M, García RH, Barrientos RJ, Fredes C, Keller AV (2020) A review of convolutional neural network applied to fruit image processing. Appl Sci 10(10):3443. https://doi.org/10.3390/app10103443

UN Department of Economic and Social Affairs (2017) World population prospects: the 2017 revision. Technical repory

Williams H, Ting C, Nejati M, Jones MH, Penhall N, Lim J, Seabright M, Bell J, Ahn HS, Scarfe A, Duke M, MacDonald B (2019) Improvements to and large-scale evaluation of a robotic kiwifruit harvester. J Field Robot 37(2):187–201. https://doi.org/10.1002/rob.21890

Chapter 9
Post-harvest Sorting and Grading

Andrew McGlone and Kerry Walsh

Abstract This chapter provides an overview of the modern sorting and grading technologies used for horticultural produce, with a focus on the commercial use of machine vision and near-infrared spectroscopy (NIRS) and the development of firmness grading methods for fruit. Prospects for future innovations are discussed, including advances in data processing algorithms and the deep imaging methods of MRI and X-ray technology. Barriers to adoption are considered and two case studies are presented illustrating the opportunities and problems encountered with commercialisation.

The high value and perishable nature of the outputs of many horticultural production systems is such that post-harvest quality assessment and management can be pivotal to maximising returns and minimising waste. Indeed, the emergence of the global trade in horticultural produce is inextricably linked to innovations in sorting and grading. Accordingly, this chapter begins with a brief history of the evolution of these post-harvest technologies.

Keywords Post-harvest · Automated sorting · Grading systems · Precision horticulture · Machine vision · Spectroscopy

9.1 The History of Sorting and Grading

The origin of 'agriculture' is thought to have begun about 10,000 years ago and is typically tied to the domestication of plants and animals, with selective harvesting of superior product (e.g., larger, sweeter fruit) resulting in initially accidental, and later deliberate, selection of genotypes suited to human use. The next steps in the development of agriculture are commonly considered to involve storage of produce

A. McGlone (✉)
New Zealand Institute of Plant & Food Research, Auckland, New Zealand
e-mail: Andrew.McGlone@plantandfood.co.nz

K. Walsh
Central Queensland University, Rockhampton, QLD, Australia

P. Newell Price et al. (eds.), *Precision Agriculture in Field Horticulture*, Progress in Precision Agriculture, https://doi.org/10.1007/978-3-031-96534-0_9

and the creation of food surpluses. The selection of produce to store would again have required manual selection, based primarily on visual or hand feel assessments.

Later, trading between population groups developed. The addition of the extra cost of transport and marketing would have provided an incentive for grading. Higher quality produce samples will always have drawn favourable attention and higher prices, while defective produce, for example overripe or diseased samples, will have been downgraded or discarded as waste. Any market enterprise depends on a positive balance between profit and loss, so the role of post-harvest sorting and grading in sustainable production will have always been clear. The technology developed to minimise losses in transport and markets, driven by a desire to maximise sale values.

Extraordinary effort was expended on the selection and transport of small quantities of fresh produce for the elite in ancient civilisations. Citrus fruit were traded across central Asia to Rome, with the oldest evidence of a lemon in Rome dated to after the late first century BC (Langgut 2017). There is also evidence that cellars were used to store fruit (apricots and melons) for aristocrats in the Western Zhou dynasty (1046–771 BC) in China (Zhouyong 2010). Writing about pepper in the first century CE, Pliny the Elder noted "Why do we like it so much? Some foods attract by sweetness, some by their appearance, but neither the pod nor the berry of pepper has anything to be said for it. We only want it for its bite – and we will go to India to get it. .... Pepper and ginger both grow wild in their native countries, and yet we value them in terms of gold and silver" (Hancock 2021).

Indeed, selected high quality and/or exotic fresh produce has been regarded as a luxury in all ages and cultures. Shakespeare described where to locate the best strawberries (Shakespeare 1599). In the classic 1987 film 'Babette's Feast', set in an austere 1870s Danish fishing village where the diet is fish and potato, Babette spends her fortune on one meal for the village of exotic foods, including papaya, figs, grapes, pineapple and pomegranate (Axel 1987). Today, modern Japanese culture exemplifies the luxury status that carefully graded and sorted produce can obtain, where fresh fruits like melons and grapes can rival quality wines as prestige gifts, as well as on price (Le 2021).

Exotic or out-of-season fresh fruit and vegetables has transitioned from a luxury to a bulk commodity with the development of bulk transportation and large-scale agriculture. For example, the transcontinental railroad across the USA enabled Californian citrus to reach a continental market from 1870 (Fig. 9.1). The increase in the volumes of production required by these markets drove innovation in produce handling and grading.

These trends in scale of production and transport rapidly accelerated, driven by population growth and rising affluence. Over half of the world's population (57% in 2021 (UN 2021)), live in cities and are dependent on the transport of fresh produce, with access to out-of-season produce at bulk commodity pricing. The global trade in fresh fruit and vegetables is through complex supply chains handling millions of tonnes of produce annually. However, horticultural produce is highly perishable and susceptible to disease and pest damage. It requires proper handling, processing, packaging, storage, transport and retailing to ensure that the fresh produce reaches

Fig. 9.1 Mechanisation of grading and packing operations was driven by increases in the scale of operation and the cost of labour. (Reproduced from Citrus 1905; Wikimedia Commons)

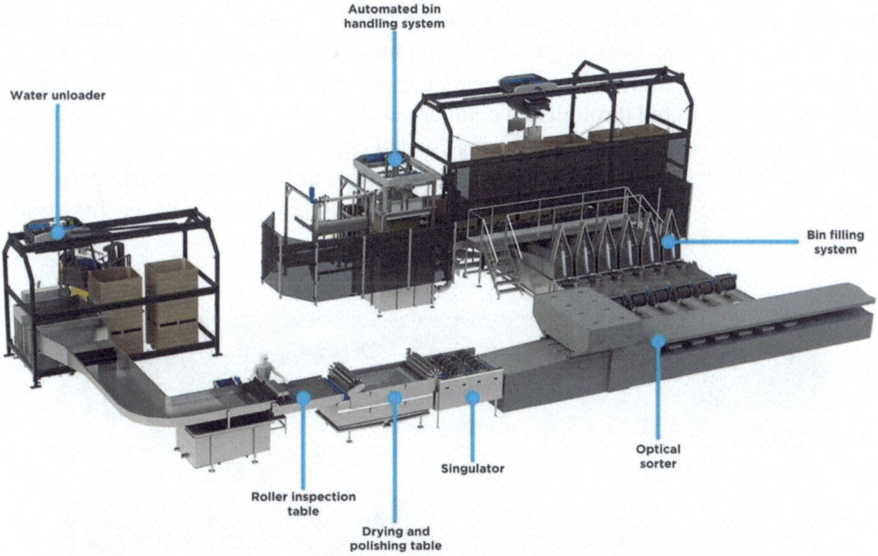

Fig. 9.2 Typical design for a modern fruit packing line. (Reproduced from WymaSolutions 2022, with permission from Wyma Engineering (NZ) Ltd)

the consumer in good condition. The global nature of trade also introduces biosecurity issues, prompting phytosanitary regulations that require produce inspection. An essential feature of the supply chains are fast and efficient mechanised packing facilities (Fig. 9.2), highly automated and equipped to grade produce based on quality attributes, segregating to waste streams any poor quality produce unfit for further processing through supply chains and/or enabling optimal differentiation for the various value chains leading to the consumer.

9.2 Grading Mechanisation and Modern Automated Scanning Systems

The earliest 'modern' mechanisation of fresh produce grading and sorting was the introduction of a moving conveyor belt or rollers, driven by motorised chains, that enabled efficient flow and presentation of produce to teams of people to grade and sort. So called 'grading stations' or 'tables' (Fig. 9.3) remain a necessary adjunct in modern packing facilities, particularly for dealing with produce problems that the automated sensor-based grading systems fail to solve or sort properly.

Produce size grading is a primary task in a packing facility, where size is an important consumer attribute and sorting is critical for optimised packing of produce into cartons. Mechanisation of size grading quickly developed with the introduction of the moving conveyor systems, such as with the invention of diverging belt systems (Fig. 9.4). In these mechanical sizing systems, pairs of diverging rods or belts are spaced such that fruit passing along would drop at a point related to fruit size. Many other mechanical grading mechanisms were also developed, such as vibrating screens or rotating drums made of a sequence of progressively larger mesh holes for produce to fall through and counterbalanced conveyor cups that allowed for weight grading. There was even a spring-loaded cup system successfully developed in New Zealand that literally threw fruit into different lanes according to their mass—it became a mainstay in the New Zealand kiwifruit industry for nearly

Fig. 9.3 Grading tables in operation. (Reproduced from Honeyfield 2014, with permission from A. Honeyfield)

Fig. 9.4 A diverging belt (ca. 1920) delivering fruit of different sizes to a table for grading before packing. (Reproduced from Pitts 1920, Museums Victoria: public domain)

20 years from invention in 1964 (Martech 2005). Mechanical systems to sort produce based on size and weight etc. are still in use today.

The first move towards electronic systems began in the 1970s with the introduction of the load cell for measurement of sample weight (Fig. 9.5). The load cell remains the standard for size grading of individual produce pieces, with typical accuracies of ±1 g. This provided a platform involving an electronically controlled system with a sensor and control of drop points that then allowed for further addition of other electronic sensors.

A watershed time for sorting automation began in the 1980s with advances in electronic camera technology and the possibility of machine vision scanning applications through digital image processing. This capacity was initially used for grading fruit on overall colour. Size sorting by machine vision, rather than weight, became possible, an option that allowed for use of wide belt conveyor systems. The development of other non-invasive technologies, notably various scanning methods and spectroscopies developed in the medical field, followed, with transfer of such technologies to the fresh produce sector. By the early 1990s there was sufficient research activity on a variety of technologies and instrumental methods relevant to assessment of internal quality of fresh produce for several review papers and specialist conferences (BARD 1994; Abbott et al. 1997; NRAES 1997; Abbott 1999).

The two main limiting factors on the adoption of non-invasive technologies, i.e. techniques that do not damage the produce, into the fruit pack-line are cost and speed. Fruit grading involves bulk assessment of relatively low-cost items. The acquisition cost for a modern grading and sorting system is in the order of USD400k per lane (e.g., see Smith 2012; adjusted for inflation). Adding technology such as a magnetic resonance imager, though an undoubtedly useful addition, is an

Fig. 9.5 A load-cell weighing bridge arrangement. The conveyor carrier unit (black plastic piece) slides across the segmented steel plate, which is attached to a load-cell (marked with an arrow). Wear tracks from the moving carrier units can be observed on the top surface of the steel plate

unlikely prospect at around USD1m. Further, the ever-present requirement to increase packline efficiency imposes a 'need for speed', and the high speeds of modern grading systems impose a primary limitation to the sensor technologies that can be applied. The practical lower speeds on modern cup conveyor systems are around 1 metre per second (m/s), in practice requiring measurement scanning times of less than around 30 milliseconds (ms) on medium-sized fruit such as apples and as little as a few ms for small fruit such as berries. The wide belt grader systems run at higher speeds, up to 5 m/s belt rates, and so scanning times must be even lower (e.g., 6 ms or less). The non-invasive assessment technologies of load cells, machine vision and near-infrared spectroscopy (NIRS) have proven viable at these speeds, becoming commercially available from the early 1970s, 1990s and 2000s, respectively (Fig. 9.6).

There is a 'law of diminishing returns' at work in the adoption of technologies into fruit grading. Assessment of the major attributes of size, weight, shape, surface defects and a number of internal gross attributes is currently covered by mechanical systems and the methods of machine vision and NIRS. The requirement for sorting on other attributes is less universal, i.e., not required for every commodity, and perhaps only at times for a given commodity. For example, the need for a grading technology to detect internal dryness defect in citrus is confined only to those seasons in which the defect occurs at a higher incidence. Thus the additional technologies tend to deliver relatively less value.

Fig. 9.6 Top panel: an eight-lane scanning system for citrus with a row of eight NIRS measurement heads (reflectance systems) immediately in front of a large machine vision system (Spectrim™ cabinet). Bottom panel: tomato colour sorting using a wide belt grader. Poorly coloured or immature fruit are identified by an optical system (yellow strip of light emitting diode (LED) lights, lower left of centre) and then tapped back to a waste stream by firing the appropriate piano key (blue levers). (Reproduced from Tomra 2022, with permission from Tomra)

Therefore, a range of other non-invasive technologies deployed in medical and other industries, and used in postharvest research, have not achieved commercial adoption in fresh produce grading. An example is nuclear magnetic resonance (NMR), which in the 1990s was being promoted for non-destructive measurement of internal fruit condition (Tollner et al. 1994; Kim et al. 1997). However, NMR is slow, with the alignment and relaxing of nuclear spins requiring typically seconds of time, and it otherwise offered no clear advantage over NIRS as a quality assessment tool. The related deep imaging technology of MRI does offer the advantage of internal spatial imaging. This capability allows detection of localised internal defects. This assessment is even slower and more expensive, although it remains a distant prospect in some circumstances as will be dicussed below. Volatile or gas sensing of produce attracts investigations from time to time, and certainly may have prospects for scanning bulk storage packages such as trays, boxes or bins, but has thus far also proved too slow to be compatible with fast on-line grading systems. Other sensing technologies, such as OCT (optical coherence tomography) for assessment of some under-the-skin defects (Li et al. 2019), also attract attention but have yet to find commercial use.

9.3 External Quality Determination by Machine Vision

Machine vision technology, using digital camera imaging, is widely adopted in fruit packhouses for measuring external properties such as size and shape, colour, blemishes and damage defects. Commercial systems consist of an illumination cabinet with multiple cameras, or camera views as achieved with mirrors, with use of various illuminating light sources. As in other application areas, in the last decade light emitting diodes (LED) have dominated as light sources for fruit grading, given their longevity, rapid warm up time, strobing capability, and availability as relatively narrow wavelength band light sources (e.g., different visible colours, UV, NIR).

Several names are used to describe this technology on a fruit packline, such as camera box, camera system, imaging cabinet and vision cabinet. In one approach, individual items of produce are carried at high rates into the cabinets on a conveyor system, typically running at 1 m/s or higher speeds. Multiple image captures are made as the samples travel though the cabinet, often whilst being rotated backwards, at low angular rates, to enable full camera views of all sides during their passage through the cabinet. In another approach, unseparated items are carried on a belt through the imaging cabinet, with only part of the surface of the fruit viewed.

Early machine vision systems provided colour grading only, based on an average fruit colour. Each generation of sorting technology since has offered increased image analysis capability, particularly around defect detection.

The need for defect detection is driven by retailers who impose specifications. In practice there is something of an ‘arms race’ with improving sorting technology leading to tightening of specifications and vice versa. As an example, a retailer specification might allow a number of small area defects with a total area of 2 cm^2, but no single defect of area greater than 1 cm^2. Another specification may accept some overall colouration but be intolerant of a colour patch associated with an insect, e.g., pink spot on mango fruit caused by scale insect. In another example, the depression associated with the perianth of the apple fruit is acceptable, but a small insect entry hole is not. Specifications can also shift with season and fruit availability, making the role of the grader operator a dynamic one.

The ability to detect a blemish or defect is determined by various technical illumination/camera factors that are difficult to tease apart as they convolve together, giving rise to the system ‘optical transfer function’. Because of these interactions,

Fig. 9.7 Visible (left) and short wave infrared (right) images of a bruised apple. (Reproduced from Grodzki 2020, with permission from Teledyne Imaging Group)

simple questions about performance limits become difficult to answer beyond such clichés as 'if you can see it easily by eye then a camera should detect it too'.

The primary limitation to detection, be that by human eye or machine vision, is the contrast between the object of interest and the background in the recorded images. Contrast can be achieved in terms of texture (spatial patterns) and/or colours (red, green and blue (RGB) differences). For example, as a first step in most packline image processing routines, the fruit image must be segmented from that of the background. Use of a conveyor cup or finger with a contrasting colour to that of the fruit will assist this step. For segmentation of a defect on the fruit surface, imaging systems can be designed to maximise contrast for a given application through choice of illumination and detection wavelengths, light source-object-camera orientations, and other techniques (Fig. 9.7).

The object to background contrast ultimately determines the practical limits of detection, as in the true positive rate. An additional consideration is the presence of other objects or defects that mimic the defect of interest, such as stems or calyxes. This issue causes false positive detections.

Image handling and processing methods of images acquired on packlines may start with either stitching images together, as a round-the-sample strip, or by rendering the image sequences into an approximation of a 3D model of the sample. Fruit size calculations can be made given knowledge of the camera to fruit distance and use of stereovision when more than one camera is employed.

Image recognition algorithms are required for defect detection. These algorithms vary in complexity (Fig. 9.8). For example, a defect might be detected using some simple rules to segment pixels with certain RGB levels, blob these pixels and accept as 'positives' blobs of a certain area. A slightly more sophisticated routine might require pre-processing of the image to obtain 'features'. The last decade, however, has seen the rise of algorithms involving convolutional neural networks, in so-called deep learning approaches (Koirala 2019).

The particular challenge involved in image processing on a packline is speed. At a conveyor speed of over 1 m/s there can be over 10 fruit assessed per camera per lane

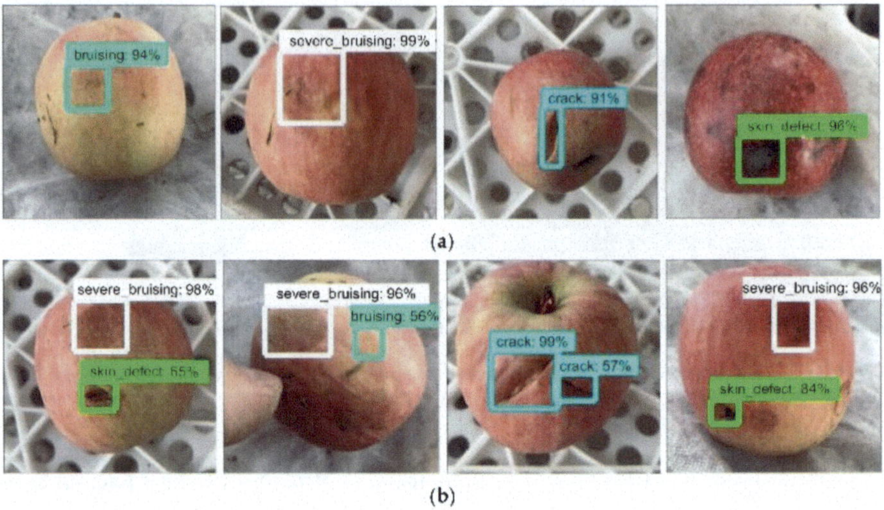

Fig. 9.8 Example of success rates of machine vision surface defect detection algorithms. (**a**) Apples with one kind of surface defect, (**b**) apples with two kinds of surface defects. (Reproduced from Hu et al. 2021; Open Access)

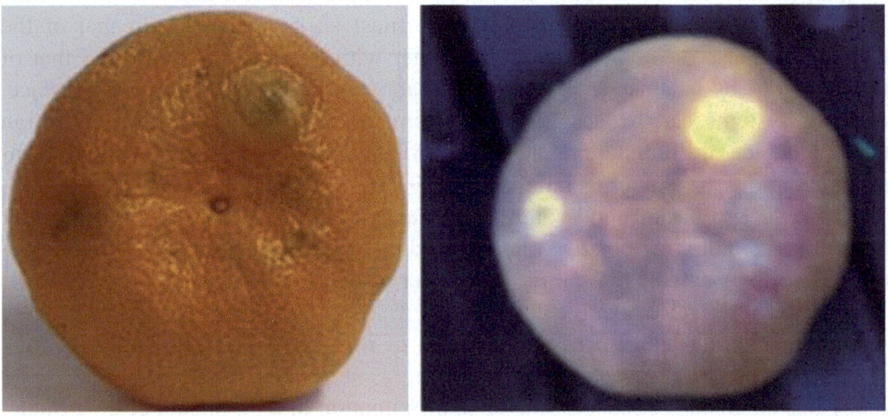

Fig. 9.9 Visible (left) and UV (right) illumination of a mandarin with clear rot symptoms, demonstrating improved contrast as a result of UV fluorescence

per second, with multiple images acquired per passing sample. Much higher rates are involved in grading smaller objects, such as berries and nuts.

The image processing challenges involved are algorithmically deep with peculiarities pertaining uniquely to each system. This means much of the intellectual property in the various commercial machine vision systems relates to the bespoke complex software code that implements the analysis.

More advanced systems allow for training of the machine vision to recognise a specified defect against specified backgrounds in the packhouse. The training involves presentation of positive and negative defect examples of the samples to the machine vision system followed by automated application of image processing and machine learning techniques to derive a suitable detection algorithm. Such 'on-the-go' training is sometimes presented as a 'quiz' that the grader operator performs by tracking the known examples through the grader. The operator will generally have some control input, such as in setting of the segregation cut-points.

In most circumstances the performance of a system can only be determined empirically, through validation trials. Such trials involve presenting samples other than those used for training to the machine vision system and observing how successful the segregations are, generally by inspection of fruit at a designated defect drop point.

In the last decade there has been greater commercial use of multispectral imaging, using discrete-waveband light sources from the UV to the NIR. This has been triggered by the availability of relatively inexpensive LED lighting available at a range of wavelengths. The discrete waveband light sources can be used for contrast enhancement between the defect and the underlying fruit surface of the sample in certain applications. For instance, UV illumination of citrus fruit results in fluorescence of clear rot affected areas, providing enhanced contrast in the image (Fig. 9.9). Similarly, there can be contrast advantages in use of NIR illumination (e.g., in detection of moisture-related defects). Defect contrast in the visible range is also possible by use of selected discrete wavebands that, for example, accentuate pigmented areas.

The fast laser scanners used for examining small produce items, such as macadamia and almond nuts, provide another example of multispectral systems. These systems consist of one or more laser beams that are rapidly scanned across the belt conveyor used to transport small items. The incident laser beam is absorbed or scattered from different samples in different ways, enabling the detection of foreign samples (e.g., dirt, stones) and/or defects (e.g., broken shells). The simultaneous use of multiple laser beams operating at different peak wavelengths results in a signal profile with often superior performance for object detection, and utility in detection of some internal features (e.g., The Nimbus BSI by Tomra, Norway; The Evolution by SATAKE, Japan).

Hyperspectral imaging involves the use of many more wavelengths, often hundreds, providing spectrally dense spatial data. A typical design for conveyor-based systems is the 'push broom' configuration, in which the object is advanced through a detected line. Received light from that line is dispersed across a 2D detector array, providing spectra for a series of segments across the line. The technique has good potential for identifying an object with a spectral fingerprint different to its background and has been widely used in at-line systems (i.e., beside the main grading system, on selected subsamples) for the detection of contaminants. However, data volumes are very high, data acquisition can be slow and the hardware itself is expensive, which are all limiting factors for use on high-speed fruit graders. Such systems can be useful for identifying wavelength choices for use in a multispectral machine vision system (Lorente et al. 2012).

The prospects for direct detection of pests and diseases on fruit relies on the visibility of the defect. Basically, if a defect is on or near the surface, and easily resolvable to the order of 1 mm in dimensions (e.g., visible with the naked eye), there is potential for detection. Detection can be improved with enhanced imaging, e.g., using different light sources to accentuate contrasts, and such enhancements will improve with time. However, many pests and diseases of biosecurity concern are small and either internal to the fruit, or located in external crevices, e.g., under the calyx, with the wash process removing objects from the visible body of the fruit. The current value proposition for use of machine vision is thus in the removal of damaged (e.g., cuts, bruises) produce that are more likely to suffer pests and diseases of biosecurity concern.

9.4 Internal Quality Determination by Near Infrared Spectroscopy

A spectrometry system involves measurement of absorption at multiple wavelengths. NIRS is a technology reasonably widely used in fruit packhouses mainly for measuring internal chemical constituents, such as dry matter content or soluble solids content (SSC), or tissue properties such as the presence of defective tissue, e.g., internal browning and rots. The term near-infrared strictly applies to a particular waveband of light, beyond the red colour end of the visible light range (750 nm) and up to 2500 nm. However, fruit grading applications typically only involve detection to 1100 nm, because of the lower cost of the silicon-based detectors that operate over the range 400–1100 nm, compared with the Indium-Gallium-Arenside (InGaAs) and Lead-Sulfide (PbS) detectors that operate to 2500 nm, and lower absorptivity by water of the wavelength range below 1100 nm. The latter feature allows greater optical pathlengths, including up to full transmission through intact fruit such as large apples. This wavelength range is referred to as SWNIR (short wave NIR). As the silicon-based detectors used in these SWNIR spectrometers are also sensitive to visible light, their design often extends to analysis of all or part of the visible range. Various alternative acronyms are used for such ranges in the literature, such as Vis-NIR or VNIR (visible to NIR range).

In a grader-based NIRS system, the sample is illuminated with a light beam containing NIR radiation. A broadband illuminating light source is normally used—for example, an incandescent halogen lamp that has its peak emission in the NIR region. The silicon spectrometers are usually quite spectrally dense, with optics built around linear sensor arrays of some hundreds of contiguous sensor elements (photodiodes). In multispectral NIR systems, as used in internal defect detection, narrower band sources such as LED are used.

The source light beam is generally used to illuminate one area or zone of the sample, with simultaneous detection of the transmitted light spectrum from another area or zone. Early generation commercial NIR grading lines employed a reflectance geometry, wherein the detector viewed the illuminated region of the fruit. In such systems most of the detected light is specular reflection, with no information on internal sample composition, or light that has diffusely scattered through only a very shallow internal portion of the fruit. The ratio of the contribution of these components can alter with change in fruit skin condition, affecting the performance of predictive models. In consequence, all commercial NIR grading systems now employ partial or full transmission geometries.

If the optical source-sensor arrangement has the illumination and detection zones close by on the same side of the sample, then the emitted light will come from only a shallow part of the fruit (i.e., only part of the sample is assessed). Other systems involve a detector zone that has a substantial angular distance from the source zone, such as at 90° (partial transmission) or 180° (full transmission). These transmission systems achieve much deeper light penetration for the detected light, although still with a sensitivity bias towards the near surface region of the detection zone (Fig. 9.10). However, they have a much lower proportion of incident light detected, and require much stronger light sources, more sensitive detectors, or longer detection times, particularly for larger fruit. Whether to use the geometry of same-side reflectance or some form of transmission for a given application depends on the nature of the fruit (e.g., the presence of the thick skin on fruit such as citrus) and/or on the defect (e.g., internal browning deep in an apple).

A downside of a full or partial transmission geometry is the low intensity of transmitted light, requiring increased integration times to improve signal to noise ratio in the detected signal. Operation at a grader rate of 10 fruit per second, typical for lane speeds of 1 m/s, limits spectral acquisition times to less than 100 m/s and involves movement of the fruit past the detector at 1 mm for every 1 m/s. The net

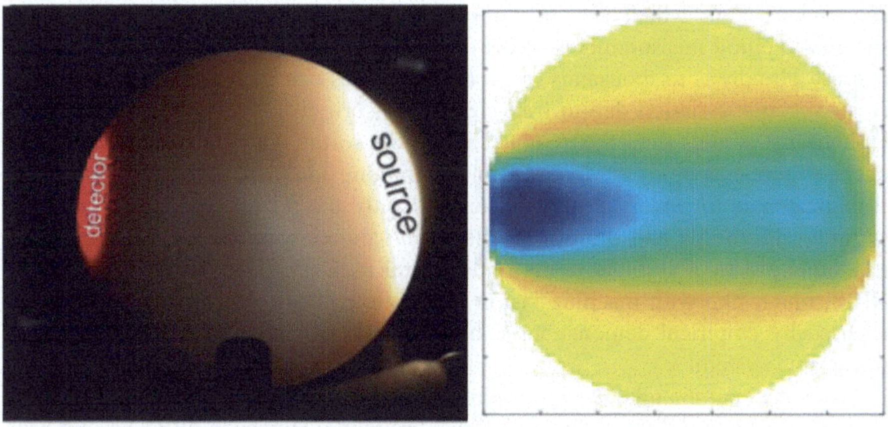

Fig. 9.10 On the left is a phantom fruit sample, a solid white Teflon ball, photographed while passing through a NIRS transmission system with the source and detection zones indicated (NB detection zone made visible by running red laser beam backwards through the detector optics). On the right is a computer simulation of the same situation, with internal light transport modelled to show the relative sensitivities of the detected light to different regions of the phantom, in equatorial cross-section, from low (yellow) to high (dark blue) sensitivity (e.g., the relative probability that a detected photon passed through a region)

result is poor spatial resolution, the detection zone blurred by the movement of the fruit on the conveyor. This is further compounded by a generally wide illumination area (e.g., centimetres in diameter) to optimise light input to the sample, and the strong scattering of light within fruit tissues—complete scattering disorientation of individual photons happens within as little as 10 mm of internal light travel. These constraints severely limit the detection of small internal defects. As a rule of thumb, a defect less than about 10% of sample volume is too small to detect, even when quite close to the surface of a fruit (Sun et al. 2020). Full sample interrogation is also not possible with a single measurement as internal sample sectors beyond the main central illumination axis will have very low interaction with the transmitted light (see Fig. 9.10). Multiple measurement stations are used to provide full sample interrogation, e.g., as used in the detection of internal browning in apple.

The acquired spectrum must be interrogated by a suitably trained and calibrated computer algorithm to deliver a prediction of the quantitative value or presence of the feature of interest. The fundamental spectral basis for most successful applications is claimed to be the existence of relevant light absorbing constituents (e.g., plant pigments, water). However, the co-existence of many interfering factors, from overlapping absorbance spectra of various constituents to changing scattering properties of the sample, means the claim is largely void of meaning. In practice, NIRS has only been reliably successful for predicting macro-constituents present at moderate percentage levels, such as SSC at >5% w/w, rather than for more minor components, such as the titratable acidity (TA) which is typically <1% w/w in most fruits (Walsh et al. 2020). Successful predictive models can only be established through careful experimentation and development of multivariate models bespoke to the application. The training process is a machine learning one, often called chemometrics, in which statistical tools are used to match paired spectral and standard method measurements of the quantity or quality of the produce feature of interest. Hundreds to thousands of produce samples can be required to adequately cover spectral variation due to fruit size, shape, origin (e.g., orchard or region) to build robust models since the training can only be secure if all likely future examples of the produce samples are included in the training samples. To say the process is complicated is to understate the matter and only the advent of modern machine learning algorithms and thorough calibration/validation processes have made the technique practically viable in the horticultural industry. The success or failure of NIRS, much like with machine vision applications, can often only be confidently asserted by empirical demonstration, by trial-and-error measurement on an actual grader line system.

NIRS is in commercial use for grading based on the attributes of SSC, dry matter content and internal defects such as internal browning. However, NIRS has not been commercially developed for detection of internal disease or detection of pest damage. This is despite some promising research on these topics, such as in the detection of fruit fly larvae in mango (Saranwong et al. 2010). This can be attributed to the small size detection limit, that in practice a defect must not only have high contrast (e.g., be strongly absorbing of NIR light compared with good tissue) but also must be at least as large as 10% by volume relative to the sample size. The diffuse nature of

NIR light transmission, coupled with the poor spatial resolution, means small defects caused by pests or diseases will be simply too diluted by the signal from the intact or good tissue. There is also the salient issue of sample orientation; if the defect position in the sample is unknown in advance, the optimal light transmission path for detection cannot be ensured. There are some interesting new NIRS prospects being researched regarding attempts to circumvent these problems, to significantly improve spatial resolution and deal with orientation issues using fast scanning multi-wavelength laser systems (Sun et al. 2020). Technologies allowing fruit orientation (https://www.roboticsplus.co.nz/robotic-apple-packing) may also help simplify defect or pest detection.

There are often claims made that NIRS can be used for firmness assessment, with suggestions that such measurement is based on change in the scattering coefficient of produce as firmness changes, or to a change in cell wall composition. Such claims have been questioned, e.g., in the review by Walsh et al. (2020). Firmness changes during fruit ripening occur concomitantly with change in a number of other attributes, such as skin pigmentation, flesh pigmentation, and starch to soluble sugar conversion, which are assessable by NIRS. With a high correlation between these attributes for a given population sharing pre-harvest and postharvest history, an apparent correlation between NIR spectra and firmness may be an indirect association. This does not matter in terms of firmness assessment as long as the attribute correlations are stable. However, any change in pre- and post-harvest condition is likely to affect such associations, reducing the predictive power of the NIRS method. For example, mango fruit held at low temperatures will show some starch to sugar conversion without skin pigmentation changes. Mango trees provided with higher nitrogen will produce fruit that apparently do not de-green although they soften during ripening. In summary, NIRS may be used in assessment of intact fruit firmness, but models are expected to lack robustness across diverse populations.

9.5 Produce Firmness Sensing by Mechanical Methods

The concept of fruit firmness is not particularly well defined in relation to produce quality (Abbott 1999). Often it is used in the sense of the touch feel of a whole produce piece, such as experienced by a consumer 'testing' produce in store by hand squeeze or finger poke. In mechanical terms, this is close to the concept of material stiffness, the measure of surface deformation under an applied load. Another firmness sense, often termed the texture, is that of mouthfeel when biting into the fruit, and includes the shearing and compressing of the tissue as it is chewed. The most popular standard industry specification for firmness is a destructive mechanical measure of such textural characteristics. It is called the penetrometer method and involves first (generally) removing the outer skin layer and pressing a metal cylinder of set diameter onto the flesh until tissue rupture occurs and the cylinder travels to a set depth. Generally, the peak force at rupture is measured. Different commodities

Fig. 9.11 The Sinclair iQ (left) and Aweta AFS (right) in-line firmness sensors. (Reproduced from Khalifa et al. 2011 (left; Open Access) and Aweta 2022 (right; with permission from Aweta))

and supply chains will have different specifications in terms of cylinder diameter, speed and depth of penetration (Abbott 1999).

Of course, the destructive penetrometer method cannot be used for on-line grading of every piece of fruit. Non-destructive approaches include methods that measure the speed of transmission of a shock wave or resonant frequency of sound emitted following tapping the fruit, or de-acceleration rate of an object striking the fruit. However, these techniques measure related but slightly or greatly different mechanical properties to that of stiffness or texture. Thus, there is the issue of equivalency of these non-invasive methods to the currently accepted penetrometer method—a truly valid and acceptably reliable method remains something of a 'holy grail' quest for packline manufacturers.

Several systems have been developed for use on packlines. Sinclair International (UK) developed a firmness tester that utilised a bellows system, used in their sticker labelling technology, to measure deceleration on contact with fruit (Khalifa et al. 2011). Because it measures at one location it was often employed in 'batteries' of multiple units, e.g., four in a row (Fig. 9.11, left). Aweta (2022) have claimed on-line firmness measurement based on an acoustic principle using resonant frequency with their AFS device (Fig. 9.11, right). This system was first developed in the early 2000s but has not become widely used, with little or no accessible information on its performance. The problem of practical and accurate on-line firmness sensing remains with continued research efforts being made towards new types of systems (Landahl and Terry 2012; Schaare et al. 2014).

9.6 Future Prospects

Advances in Data Processing

Significant advances in data processing are continuing to occur in the produce grading space, particularly in machine vision technology. This is both in terms of hardware allowing edge computing, and in terms of image processing techniques, with development of lighter weight artificial intelligence (AI) systems. These advances are driven by demand in other settings, e.g., vision processing in autonomous vehicle and security applications, with spin-off adoption into the fruit grading sector.

For example, a decade ago most machine vision grading systems would have employed manually defined ('handcrafted') parameters on features such as intensity, colour space, shape and texture histogram of oriented gradients (HOG), local binary patterns (LBP) or Haar-like features (digital image features used in object recognition), with creation of 'rules' for assessment of fruit characters. However, these 'handcrafted' approaches can fail in use with fruit produced under other conditions (e.g., different cultivars, growing conditions, lighting conditions).

The adoption of supervised deep learning approaches was initially constrained by the need to annotate large sets of images for the purpose of training the network. This barrier has been removed by the availability of large public annotated image sets, the use of transfer learning which enables an established model to be used as a base in a new application, and the availability of easy to use graphical annotation tools such as LabelImg (https://github.com/tzutalin/labelImg). Further, deep learning frameworks such as Faster Regional Convolutional Neural Network (Faster R-CNN), Single Shot multibox Detector (SSD), and You Only Look Once (YOLO) have the capacity to parse the commonly used PASCAL-VOC annotation format (Koirala et al. 2019). As a consequence, the use of supervised deep learning approaches has become common in recent years.

NIRS is a heavy user of AI algorithms, particularly the subset of supervised machine learning methods referred to earlier as chemometrics. The ever-present desire for more analysis capability with the technology, be that in prediction of unique or novel compositional parameters or just with improving accuracy and precision of existing predictions, means advances with new data analysis methods is a continuous research activity. As with machine vision sorting, large datasets are required for use of the modern deep learning neural network methods. However, the idiosyncratic nature of the technology hardware, where instrument specifics for any particular installation can be troublesome (e.g., nuances in the sources and optic components), means the large dataset training problem is not so easily overcome as it is with machine vision—there is very active research into this topic (e.g., Wohlers 2022).

X-ray Technology

Traditional X-ray imaging relies on visualisation of the attenuation of X-rays through an object. This is a function of density and path length. Unfortunately, fruit do not present a uniform thickness and the range of densities of inclusions (from skin, flesh, seed to insects) is similar, leading to relatively poor visualisation (Fig. 9.12) except for large voids. Consequently, X-ray technology is not widely used in the horticultural industry beyond detection of foreign bodies, such as metal objects, and large internal cavities. An often quoted horticultural success for X-ray technology has been for hollow heart in potatoes, which dates as far back as 1937 for

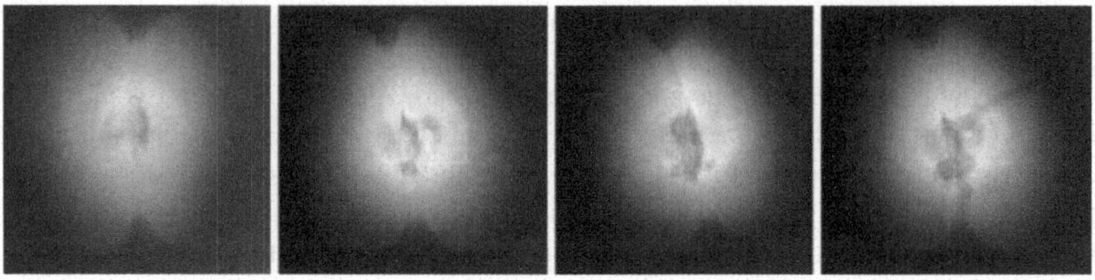

Fig. 9.12 X-ray images of codling moth infestation in an apple, age of infestation varying from left to right over 17 days from inoculation with eggs. (Reproduced from Haff and Toyofuku 2008, with permission from Springer Nature)

first investigations and in automated commercial form since probably around the 1980s (Abbott 1999). The technology is in widespread use in the fresh-cut produce and consumer pack space, where there is zero tolerance for foreign bodies such as stones or, as in recent times, metal objects such as needles in fruit (Davidson 2018).

Two main issues appear to combine to limit application of the technology in horticulture: (1) the lack of defect contrast in biological material such as fruit, and (2) the low sensitivity and high cost of the necessary digital sensors. There is also an issue of safety, X-rays being ionising radiation and thus potentially hazardous to human health, which brings extra regulatory costs into play. As mentioned above, the lack of contrast is a result of the similar material densities of fruit tissue, including seeds, and internal insect infestation. Open cavities present strong contrast, air being obviously much less dense than tissue, and so detection of defects such as insect tunnels is possible and has been demonstrated in research applications using highly sensitive photographic plates (Haff and Slaughter 2004). However, current digital sensor systems for X-rays are not nearly as sensitive as standard X-ray plates (Haff and Toyofuku 2008).

There has been some promising recent research activity towards advancing X-ray computer tomography (CT) for applications in horticulture (Van de Looverbosch et al. 2020). In its full form, X-ray CT requires the measurements of X-ray projections at a multiple of angular positions around the sample, with reconstruction algorithms then employed to render a 3D image of the internal structure. Higher value applications, such as in medicine or high-end industrial research, can accommodate the low acquisition speeds or the high hardware costs that are involved in standard approaches involving either precise sample rotation or large multiple source-sensor arrays. The suggestion from the recent work, and particularly an associated patent (Van Dael et al. 2015), is that costs could be substantially reduced and speed greatly improved by using prior knowledge, about shape and expected internal content, to enable reconstruction from far fewer projections. This could perhaps be just from line scan images assembled from the linear passage of a sample through an X-ray projection system. The rendered 3D internal structures would necessarily be quite crude, compared with standard medical X-ray CT, but might be sufficient for the purposes of detecting pests and diseases.

Magnetic Resonance Imaging

Magnetic resonance imaging (MRI) has seen tremendous growth in the medical field, driven by ever-increasing levels of information available in using the technique. In Australia, between 1998 and 2017 the number of MRI units in hospitals and surgeries increased from 38 to 508 with it now available in regional centres not just major city hospitals (Parliament of Australia 2018). MRI deploys magnetic fields and radio waves and is particularly useful for imaging soft tissues.

MRI has been used on fresh produce but only for research purposes. Typical applications of the technology with fresh produce use a strong magnetic field to flip proton (H) nuclei, with released radio waves used to capture information primarily about water mobility. It can produce strikingly detailed images of the internal structure of fruit and vegetables (e.g., Fig. 9.13, left), much superior to standard X-ray imaging. Detection of internal pest and disease issues with high accuracy is possible (e.g., Fig. 9.13, right). Unfortunately, the technique is currently slow, expensive and has some risk involved in use of high-strength magnetic fields.

However, there are some promising developments occurring that give a little optimism about MRI application to produce inspection in the future, particularly at-line. Aspect Imaging (Shodam, Israel), a claimed leader in medical MRI, have been pushing implementation of the technology towards industrial use in the fresh produce sector. They claim a point of difference in that their equipment has a low magnetic field outside the unit (e.g., the magstripe on credit cards will not be wiped clean). They have been exploring the agricultural market, developing a demonstration unit in 2015 that delivered a single slice image of citrus fruit at a 2 s per fruit rate.

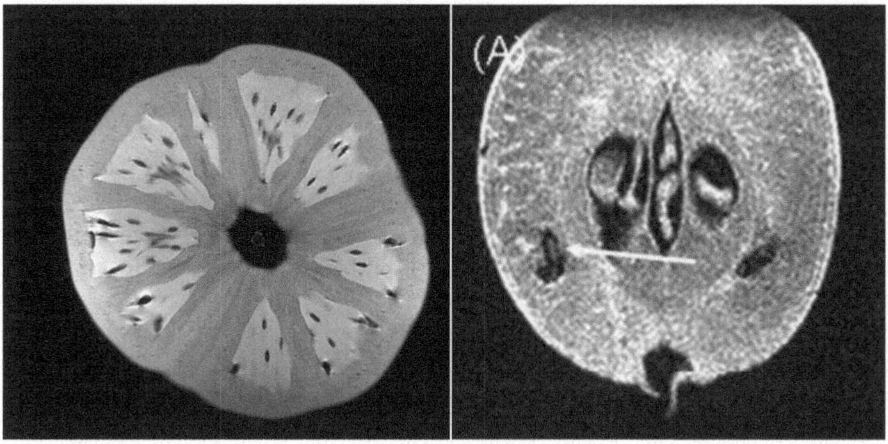

Fig. 9.13 Left panel: Magnetic resonance image of tomato fruit. (Reproduced from Ellison 2022, with permission from Andy Ellison). Right panel: Magnetic resonance image of peach fruit moth larvae in fruit. (Reproduced from Haishi et al. 2011; Open Access)

It is unclear whether this has subsequently been developed beyond that simple proof of concept. A full 3D image would be required for complete assessment of a fruit, rather than a single 'slice' image.

The technology may never be fast enough for grading line applications, there being a physical limit to the speed with which magnetic fields can be adjusted and controlled, but slower at-line applications may be possible in the future by taking advantage of efficiencies if tray or box lots of fruit can be examined at one time. It will likely remain relatively expensive, almost definitely out of reach in terms of single packhouse operations, but as an at-line system it could be easily shareable across an industry where harvest season timings vary greatly by locality and crop, or for intermittent usage tied to assessment of internal defects or infestation.

9.7 Barriers to Adoption

As noted earlier, there is a 'law of diminishing returns' to be applied to the addition of sensor technologies to the packline. Indeed, one major grader manufacturer (pers. comm.) believes that the post-harvest grading space will be essentially 'saturated' in less than 20 years in terms of practical technology. Development and application of scanning technologies is expensive, and thus an economy of scale, i.e., a large application need, is necessary to support development and commercial deployment. The result is that investment in development of such technology tends to be focused on large value commodities in the first instance. For example, in Australia, the larger value commodity industries, such as the almond and citrus industries, are now well advanced in the use of grading and sorting technologies. By comparison the smaller value Australian mango industry employs weight grading only, with use of machine vision colour and defect grading only employed as a by-product of using a grading line also used for another commodity, typically citrus.

This need for scale in post-harvest grading technology can impact on land use optimisation by creating a barrier to diversification of crops. Smaller-scale, localised crop diversity that may be desirable from a sustainable land use perspective may not be compatible with the need to raise the necessary investment capital for sorting and grading infrastructure. Trying to address this by scaling with others across distant localities may in turn create severe logistical problems such as coordinating crop timing and transportation.

There is also the recently emergent prospect of in-field sensing technologies that may circumvent much of the need for post-harvest grading technology. The refrain heard from some in the industry is that post-harvest grading is just the 'ambulance at the bottom of the cliff', that it would be far better, more sustainable, to intervene earlier pre-harvest and prevent problem defects occurring at source. In-field orchard sensing is obviously much more difficult, a confounding inversion of the grading situation with the sensors to be taken to the fruit and with the added complexities arising with varying outdoor environmental conditions. It will also depend on fundamental knowledge about the generation of fruit attributes and/or defects, the prediction of outcomes in terms of fruit growth. This is a very active research area, often going by terms such as digital horticulture and orchard digital twin, and involves use of orchard scanning technologies such as lidar and machine vision (Edwards and Moghadam 2019).

9.8 Concluding Remarks

The role of grading and sorting in the fresh produce supply chain has been examined, with a focus on describing the main sensing technologies in use and how they work to detect and segregate on quality attributes. Technologies for sorting by weight using load cells, by external attributes (e.g., shape, blemishes, colour) using machine vision and on internal attributes by NIRS (e.g., dry matter, SSC, defects) are 'mature'. However, technological advances continue, most significantly with respect to software and the modern data processing algorithms developed in computer vision. Technological gaps still exist, particularly the need for accurate flesh firmness sensing and for deep imaging methods that can accurately detect small internal defects. Currently, the single largest barrier to greater use of grading technologies appears to be the high capital costs involved, restricting use to only large value and volume commodities.

Two case studies are presented below that highlight some of the practical difficulties faced in generating sufficient value to justify the costs involved. The application of the more sophisticated—and expensive—technologies such as MRI may see adoption first in at-line applications in support of biosecurity applications. It is also anticipated that the development of practical in-field sensing technologies will continue, significantly reducing the need for post-harvest grading by enabling orchard interventions to optimise final production quality and so reduce and/or prevent waste at source.

9.9 Case Studies

The following two case studies, involving NIRS, illustrate the industrial context affecting grading and sorting operations.

Adoption in Australia of Technology for Grading Fruit by SSC

NIRS has been extensively used in agricultural industries, particularly forage, grains, dairy and sugar, since the 1970s. Application of this technology to grading of fruit for internal carbohydrate levels began decades later, starting commercially in 2000 in Australia, although Japan had several commercial applications in the 1990s. Thus, although the technology was not mature for this application, technical advice was available in country and appropriate sensor hardware was commercially available. These factors supported development of NIRS sorting on a fruit packline by an Australian packline manufacturer. However, initial demand by packhouses was not rapid, due to cost (approximately USD30 000 per lane at the time), relative complexity of the technology (need for regular calibration) and uncertainty as to the market demand for sorted fruit. Consideration on how to extract value from the technology was required.

Consumer focus group research was undertaken by a fruit marketing group to identify commodities with "flavour gaps", i.e. a difference between taste expectation and experience (R. Gray, OneHarvest P/L, pers. comm.). Flavour gaps were identified for melons (in all seasons), stone fruit (in early season) and mangoes (in all seasons). Exclusive access to one brand of NIRS grading equipment was negotiated and a "Guaranteed Sweet" brand was launched. Contracts were offered to melon and stone fruit growers over a geographic range to provide supply security and seasonal spread. The technology was offered without charge to the packhouse, and an above market price was paid for fruit above a specified SSC level. The grower was free to market remaining fruit via channels of their choice.

However, several factors stymied the success of this model. Many established growers were locked into existing marketing arrangements. Also, while growers believed they grew consistently sweet fruit, in practice pack out rates on produce above the sweetness specification were lower than expected. Pack out rates also varied with time across the season. Being unable to guarantee supply across weeks created marketing problems for the "Guaranteed Sweet" brand. A high proportion of under specification, poor eating quality fruit created marketing problems for the grower. NIRS grading also created logistical issues to packhouse operation. The NIRS sorting technology was offered without charge, but costs were incurred in integrating the technology into the electronics of the packline. The effort of calibration for the NIRS technology was higher than packhouse operators were accustomed to providing for equipment, especially for the melon application. General shed operations were also complicated, e.g., 6 size classes became 12 lines when size classes were segregated to two SSC bands. Postharvest issues also emerged, in that higher SSC fruit were generally more mature than low SSC fruit, and so ripened at a higher rate and required different marketing channels.

The 'Guaranteed Sweetness' labelled fruit were marketed to a major retailer with a price premium over normal supply. The higher price was supported by in-store taste testing and branding of the fruit with the 'Guaranteed Sweetness' label. This favourable arrangement was dented by two issues. Firstly, as sorting specifications on SSC were maintained, the volume of available product varied weekly. Store and shelf space was lost in low volume weeks, which proved difficult to recover. Also, the retailer dis-allowed labelling other than producer name and variety after 1 year. With unbranded higher class premium fruit marketed alongside standard fruit, sales suffered.

Without creation of value, there was no reward for NIRS technology implementation in the packhouse. Subsequently the technology was sold as an accessory to the packline, without exclusivity of marketing arrangement, with value found in several directions. In one scenario, an operator sorting and marketing fruit for a number of growers used the technology to reward suppliers of above specification fruit and penalize suppliers of below specification fruit. The reward system encouraged agronomic and varietal selection to improve fruit quality, which resulted in increased market share. In another scenario, NIRS sorting was adopted in context of a new cultivar which was under a closed loop marketing arrangement as part of an overall quality control programme to create confidence in the product.

Thus, uptake of a new sorting technology can only occur where product value is created beyond the cost of implementing the technology, which requires consumer appreciation of the selected attribute, an agronomic programme able to consistently deliver the product and a marketing programme able to achieve an increased value.

Grading Kiwifruit by Dry Matter Content Using NIRS

Research interest in fruit grading by NIRS methods began in New Zealand during the mid-1990s. The initial focus was kiwifruit and dry matter content (DMC) prediction. Subsequent publication of successful research reports (McGlone and Kawano 1998; Osborne et al. 1999) led directly to commercial manufacturing interest in the technology. In a striking timeline parallel with Australian NIRS developments, a manufacturing and marketing company, Taste Technology Ltd., was subsequently formed to commercialise the technology in 2000 (TasteTech 2000) and was first launched at a USA trade show in October that year (NZHerald 2000). Not surprisingly then, the technology had early successes in the North American citrus and apple industries. The initial marketing approach was similar to the Australia initiatives too, set around the concept of uniquely labelling the graded fruit, using the marketing moniker 'TasteMark' to signify guaranteed fruit quality.

Despite the early New Zealand research focus being exclusively on kiwifruit, it would be more than a decade before value-adding commercial opportunities with the technology would arise for the New Zealand kiwifruit industry. The New Zealand kiwifruit industry operates internationally under a single exporter, the marketing company Zespri International Ltd. The exporter judges grower supplied fruit using a quality assessment programme that assigns a Taste Zespri™ Grade (TZG) based on the average DMC of submitted populations (Zespri 2017). Minimum taste standards (MTS) are set by cultivar and higher payouts are made for higher TZG taste bands. The payment scheme is largely an incentive one, encouraging the production of better tasting fruit through the proven link between high DMC and consumer taste (Harker et al. 2009). Achieving the MTS was not generally a problem for *Actinidia chinensis* var. *deliciosa* 'Hayward' (marketed as Zespri™ Green Kiwifruit) and *A. chinensis* var. *chinensis* 'Hort16A' (marketed as Zespri™ Gold Kiwifruit) that dominated exports out of New Zealand until 2010. The devastating arrival that year of a bacterial vine disease, *Pseudomonas syringae* pv. *actinidiae* (Psa), meant the effective end of the Zespri™ Gold Kiwifruit cultivar (Woodcock 2016). Fortunately, the industry had a newly emergent gold cultivar available as a replacement, *A. chinensis* var. *chinensis* 'Zesy002' (marketed as Zespri™ SunGold Kiwifruit), with good resistance to Psa. By 2015 the industry was recovering remarkably well with export revenues returning to pre-Psa levels.

However, issues around fruit taste quality became increasingly apparent with the new cultivar. The vines were much higher yielding, potentially compromising overall DMC quality, and had to be very well managed to prevent high incidences of poor tasting fruit. NIRS grading and sorting on DMC became a commercial possibility then as increasingly large numbers of fruit populations failed the MTS. High DMC subpopulations of the fruit, exceeding the MTS thresholds, could be sorted and submitted for export. Packing facilities offering NIRS sorters increased from two in 2015 to more than ten separate facilities in 2018, with many millions of individual fruit examined and graded for export (McGlone and Wohlers 2016). Without the introduction of the NIRS grading technology, all fruit consignments failing the MTS would be dumped to waste, an unsustainable production outcome for the growers concerned.

Despite the success, the New Zealand industry has still not fully embraced the NIRS grading technology. There remain ongoing concerns about the accuracy and whether the costs, in the grading and sorting operation, outweigh the benefits in the value of the recovered fruit for export. Auditing data from NIRS systems used during the 2016 harvest season revealed only a moderate average sensitivity of 75%, with a false positive rate of about 15%, for correct prediction of low DMC fruit in grading and sorting a population (McGlone and Wohlers 2016). In addition, there were significant performance differences, on any one grader, between different sorted consignments (Fig. 9.14). It was clear then that the NIRS technology was

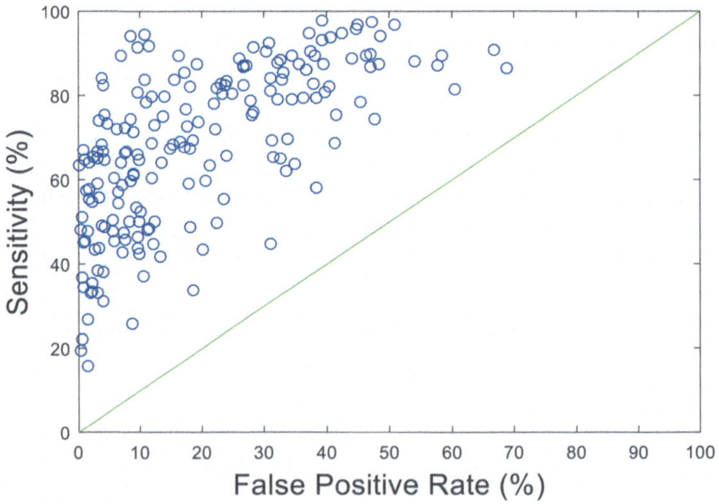

Fig. 9.14 Sensitivity and False Positive Rates estimated from data accumulated on 92 independent grower lines segregated on dry matter content at one near-infrared spectroscopy sorting facility during the 2016 New Zealand kiwifruit harvest season. (McGlone and Wohlers 2016)

underachieving because of inaccuracies, sometimes by a considerable margin. Kiwifruit industry interest in the technology continues, the benefits are obvious from a quality control perspective, and research initiatives are underway to improve the capability and performance of the systems available (Zespri International 2022, pers. comm.).

Competing Interests Kerry Walsh has received funding from MAF Roda and Felix Instruments. Andrew McGlone has received funding from various predecessors to Tomra Food and Zespri International Ltd.

References

Abbott J (1999) Quality measurement of fruits and vegetables. Postharvest Biol Technol 15:207–225. https://doi.org/10.1016/S0925-5214(98)00086-6

Abbott J, Lu R, Upchurch B, Stroshine R (1997) Technologies for non-destructive quality evaluation of fruits and vegetables. Hortic Rev 20:1–120. https://doi.org/10.1002/9780470650646.ch1

Aweta (2022) Aweta AFS. https://www.aweta.com/en/produce/mango. Accessed Jun 2025

Axel G (1987) Babbet's feast. Wikipedia. https://en.wikipedia.org/wiki/Babette%27s_Feast. Accessed Jun 2025

BARD (1994) Nondestructive technologies for quality evaluation of fruits and vegetables. Proceedings of the international workshop funded by BARD, Spokane, June 1993. Publication 05-94. ASAE, St Joseph

Zhouyong S (2010) Archeologists find 3,000-year-old fruit cellar. http://kaogu.cssn.cn/ywb/news_history/history_academic_activities/201011/t20101122_3920276.shtml. Accessed Jun 2025

Citrus (1905) Citrus packing house. https://commons.wikimedia.org/wiki/File:Interior_of_citrus_packing_house,_Ontario,_ca.1905_%28CHS-1677%29.jpg, and https://commons.wikimedia.org/wiki/File:Interior_of_packing_house,_probably_Covina,_California_%28CHS-1347%29.jpg. Accessed Jun 2025

Davidson H (2018) Strawberry needle scare: strict new export rules put in place. The Guardian. https://www.theguardian.com/world/2018/sep/18/strawberry-needle-scare-farmers-install-metal-detectors-and-lay-off-staff. Accessed Jun 2025

Edwards EJ, Moghadam P (2019) Intelligent systems for commercial application in perennial horticulture. Proceedings 36(1). https://doi.org/10.3390/proceedings2019036059

Ellison A (2022) Inside insides—magnetic resonance imaging of foods. https://insideinsides.blogspot.com/p/who-am-i.html. Accessed Jun 2025

Grodzki (2020) Imaging inside out: SWIR for apples. https://www.wileyindustrynews.com/en/news/imaging-inside-out-swir-apples. Accessed Jun 2025

Haff RP, Slaughter DC (2004) Real-time X-ray inspection of wheat for infestation by the granary weevil, Sitophilus granarius (L.). Trans Am Soc Agric Biol Eng 47(2):531–537. https://doi.org/10.13031/2013.16022

Haff RP, Toyofuku N (2008) X-ray detection of defects and contaminants in the food industry. Sens & Instrumen Food Qual 2:262–273. https://doi.org/10.1007/s11694-008-9059-8

Haishi T, Koizumi H, Arai T, Koizumi M, Kano H (2011) Rapid detection of infestation of apple fruits by the peach fruit moth, Carposina sasakii Matsumura, larvae using a 0.2-T dedicated magnetic resonance imaging apparatus. Appl Magn Reson 41:1–18. https://doi.org/10.1007/s00723-011-0222-8

Hancock J (2021) The Eastern Trade Network of Ancient Rome, World History Encyclopedia. https://www.worldhistory.org/article/1761/the-eastern-trade-network-of-ancient-rome/. Accessed Jun 2025

Harker FR, Carr BT, Lenjo M, MacRae EA, Wismer WV, Marsh KB, Williams M, White A, Lund CM, Walker SB et al (2009) Consumer liking for kiwifruit flavour: a meta-analysis of five studies on fruit quality. Food Qual Prefer 20(1):30–41

Honeyfield A (2014) Kiwifruit being packed (Quiltingorchardist). http://anorchardistquilting.blogspot.com/2014/05/kiwifruit-being-packed.html. Accessed Jun 2025

Hu G, Zhang E, Zhou J, Zhao J, Gao Z, Sugirbay A, Jin H, Zhang S, Chen J (2021) Infield apple detection and grading based on multi-feature fusion. Horticulturae 7(9). https://doi.org/10.3390/horticulturae7090276

Khalifa E, Komarizadeh M, Tousi B (2011) Usage of fruit response to both force and forced vibration applied to assess fruit firmness – a review. Aust J Crop Sci 5:516–522

Kim S, Chen P, McCarthy M (1997) NMR sensor for fruit internal quality evaluation, Publication NRAES-97. Northeast Regional Agricultural Engineering Service, Ithaca

Koirala A, Walsh KB, Wang Z, McCarthy C (2019) Deep learning for real-time fruit detection and orchard yield estimation – benchmarking of 'MangoYOLO-512'. Precis Agric. https://doi.org/10.1007/s11119-019-09642-0

Landahl S, Terry LA (2012) Avocado firmness monitoring with values obtained by means of laser Doppler Vibrometry. 4th international conference on postharvest unlimited. 2012 IV international conference postharvest unlimited 2011 945, pp 239–245

Langgut D (2017) The citrus route revealed: from Southeast Asia into the Mediterranean. HortScience 52(6):814–822

Le K (2021) Why is fruit in Japan so Expensive? https://www.tsunagujapan.com/why-is-fruit-in-japan-so-expensive/. Accessed Jun 2025

Li M, Landahl S, East AR, Verboven P, Terry LA (2019) Optical coherence tomography—a review of the opportunities and challenges for postharvest quality evaluation. Postharvest Biol Technol 150:9–18

Lorente D, Aleixos N, Gomez-Sanchis J, Cubero S, Garcia-Navarrete OL, Blasco J (2012) Recent advances and applications of hyperspectral imaging for fruit and vegetable quality assessment. Food Bioprocess Technol 5(4):1121–1142. https://doi.org/10.1007/s11947-011-0725-1

Martech (2005) New Zealand's drive for premium quality fruit and vegetables has led to high technology grading systems that lead the world. Martech Consulting Group. http://martech.co.nz/images/10packhouse.pdf. Accessed Jun 2025

McGlone VA, Kawano S (1998) Firmness, dry matter and soluble solids of postharvest kiwifruit by NIR spectroscopy. Postharvest Biol Technol 13:131–141

McGlone V, Wohlers M (2016) NIR sorting of gold kiwifruit during the 2016 season. A Plant & Food Research report prepared for Zespri Group Limited. Milestone No. 68970. Contract No. 33475. SPTS No. 13971

NRAES (1997) Sensors for non-destructive testing. Proceedings from the sensors for non-destructive testing international conference, Florida, February 1997. Publication NRAES-97, Northeast Regional Agricultural Engineering Service, Ithaca

NZHerald (2000) Kiwis take high tech fruit grader to US. https://www.nzherald.co.nz/nz/kiwis-take-high-tech-fruit-grader-to-us/TNV7B5PZVVRB3KJPMGVGPHGPDA/. Accessed Jun 2025

Osborne SD, Künnemeyer R, Jordan RB (1999) A low-cost system for the grading of kiwifruit. J Near Infrared Spectrosc 7(1):9–15

Parliament of Australia (2018) Use of MRI in Australia (Chapter 3 in "Availability and accessibility of diagnostic imaging equipment around Australia", Commonwealth of Australia). https://www.aph.gov.au/Parliamentary_Business/Committees/Senate/Community_Affairs/Diagnosticimaging/Report/c03. Accessed Jun 2025

Pitts (1920) The 'lightning' fruit grader. https://collections.museumsvictoria.com.au/items/766625. Accessed Jun 2025

Saranwong S, Thanapase W, Suttiwijitpukdee N, Rittiron R, Kasemsumran S, Kawano S (2010) Applying near infrared spectroscopy to the detection of fruit fly eggs and larvae in intact fruit. J Near Infrared Spectrosc 18(4):271–280. https://doi.org/10.1255/jnirs.886

Schaare PN, McDonald RM, McGlone VA (2014) System and method for determining a property of an object, and a valve. Patent No WO2014007660

Shakespeare W (1599) Henry V Act 1. quotetab https://www.quotetab.com/quote/by-william-shakespeare/the-strawberry-grows-underneath-the-nettle-and-wholesome-berries-thrive-and-ripen?source=berries. Accessed Jun 2025

Smith C (2012) Another feather in the cap for Compac. NZ Foodtechnology. https://www.foodtechnology.co.nz/content/another-feather-in-the-cap-for-compac/

Sun J, Kunnemeyer R, McGlone A, Tomer N, Sharrock K (2020) A spatially resolved transmittance spectroscopy system for detecting internal rots in onions. Postharvest Biol Technol 163:11141. https://doi.org/10.1016/j.postharvbio.2020.111141

TasteTech (2000) Company details. https://www.companiesnz.com/company/1060151/tastemark-limited. Accessed Jun 2025

Tollner H, Affeldt A, Brown GK, Chen P, Galili N, Haugh CG, Notea A, Sarig Y, Schatzki T, Shmulevich I, Zion B (1994) Nondestructive detection of interior voids, foreign inclusions and pests. Proceedings of the international BARD workshop, Spokane, June 1993. Publication 05-94, ASAE, St Joseph, MI, USA

Tomra (2022) Sentinel-II Food Sorting Machine. https://www.tomra.com/food/machines/sentinel-ii. Accessed Jun 2025

UN (2021) World urban population. https://statisticstimes.com/demographics/world-urban-population.php. Accessed Jun 2025

Van Dael M, Verboven P, Nicolai B, Dhaene J, Van Hoorebeke L, Sijebers J 2015. WO 2016/146703 A1. Automated Quality Control and Selection, Belgium

Van de Looverbosch T, Bhuiyan M, Verboven P, Dierick M, Van Loo D, De Beenbower J, Sijber J, Nicolai B (2020) Nondestructive internal quality inspection of pear fruit by X-ray CT using machine learning. Food Control 113:107170. https://doi.org/10.1016/j.foodcont.2020.107170

Walsh K, Blasco J, Zude-Sassec M, Sun X (2020) Visible-NIR 'point' spectroscopy in postharvest fruit and vegetable assessment: the science behind three decades of commercial use. Postharvest Biol Technol 163:11140. https://doi.org/10.1016/j.postharvbio.2020.111246

Wohlers M (2022) Augmenting NIR spectra in deep regression to improve calibration. PhD Thesis, University of Waikato, NZ (in preparation)

Woodcock SD (2016) A review of research and development undertaken on PSA. Kiwifruit Vine Health (KVH) https://kvh.org.nz/vdb/document/103504. Accessed Jun 2025

WymaSolutions (2022) Fruit pack-house solutions. https://www.wymasolutions.com/line-solutions/pack-house-fruit-growers. Accessed Jun 2025

Zespri (2017) Taste Zespri Grade. https://www.nzkgi.org.nz/wp-content/uploads/2017/05/NZKGI-Factsheet_FS-3-TZG.pdf. Accessed Jun 2025

Chapter 10
Six Elements of Data-Driven Precision Horticulture

Maryam Alavi, Roger Robson-Williams, Harris Lin, Daniel Bentall, and Linley Jesson

Abstract The use of data and analytics in precision horticulture can increase yield and quality, reduce waste, and improve understanding of consumer preferences. Advances in data science are helping overcome limitations of data collection, storage, and processing, which have been obstacles to the application of data-driven approaches to decision making. However, these advances are not necessarily applied to precision horticulture because stakeholders in horticultural value chains may not be aware of how the data technologies can add value. Ambiguity in guidelines around data licensing negatively impact data owners' trust in how their data will be used and how they can retain control over it. These challenges, in addition to the complex nature of data integration across various sources, make creation of integrated data products cumbersome and costly. In this chapter, we identify six key phases that should be considered for data-driven horticulture to ensure benefits for stakeholders, address data licensing concerns, and reduce cost of data reusage. As the trend in use of big data for smart horticulture continues to grow, taking careful consideration of the six elements of data-driven precision will enhance the horticulture industry's ability to employ big data technologies for value and impact creation across the value chain.

Keywords Data-driven horticulture · Precision horticulture · Data science · Data sovereignty · Data analysis

10.1 Introduction

Precision horticulture depends upon data. This can be acquired from a range of sources including proximal sensors, such as soil moisture probes, and remote sensors such as satellite-mounted spectrophotometers. Analysis and integration of these data can improve decision making about inputs such as fertiliser and water, assist with

M. Alavi (✉) · R. Robson-Williams · H. Lin · D. Bentall · L. Jesson
The New Zealand Institute for Plant and Food Research, Private Bag 92169, Auckland Mail Centre, Auckland, New Zealand
e-mail: Maryam.Alavi@plantandfood.co.nz

P. Newell Price et al. (eds.), *Precision Agriculture in Field Horticulture*, Progress in Precision Agriculture, https://doi.org/10.1007/978-3-031-96534-0_10

estimation of parameters such as yield and maturity estimation, and enable improvements to post-harvest management. As such, data can play an important part in increasing resource use efficiency, yield, and quality of horticultural produce. These are the goals of data-driven horticulture, i.e. the use of sensing technology and data analytics to inform management decisions and interventions in a horticultural value chain.

The benefits of data-driven precision horticulture can accrue at a range of scales from individual businesses to whole sectors. Just as the collection and analysis of data from millions of personal fitness trackers can provide new insights for improving human health (Codella et al. 2018), the smart use of data and analytics in precision horticulture can deliver benefits for whole sectors of the food system, through an increased understanding of the effects of management, genetics, and environmental conditions across the whole horticulture value chain.

The benefits of data for 'smart farming' (Plazas et al. 2019; Arya 2021), sustainable farm management (Sarker et al. 2019), detection and classification of pests (Barbedo 2020), and post-harvesting systems (Olayinka et al. 2022) are now well-established. These examples, along with many others, show that integration and sharing of data throughout the value chain has considerable potential to enable increased yield and quality, reduced waste, and improved understanding of consumer preferences.

However, the application of data integration and big data analytics is not straightforward in precision horticulture. The cost and complexity of data collection, integration, and analysis can rapidly become prohibitive and outweigh the value that it brings. Additionally, there is a lack of clear guidance on data sharing, data ownership, and value sharing. A lack of clarity on data licensing also acts as an obstacle to data integration from different parts of the value chain, yet there is evidence that the more transparent the licensing, the greater the trust by data providers (Wiseman et al. 2019).

Common 'core principles' or data use agreements are often used to increase transparency and reduce complexity of data use licencing agreements e.g., Core Principles — Ag Data Transparent.[1] The findability, accessibility, interoperability, and reuse of digital assets—referred to collectively by the acronym FAIR—were published in 2016 by Wilkinson et al. and are considered the gold standard to enable data integration across an industry. Despite this, surveys of growers and farmers have shown that sharing of data has a long way to go for acceptance. The reasons for this include fear of increased regulation, concerns about privacy or trade secrets as well as lack of trust in the downstream users of these data (Brini 2020; Wiseman et al. 2019). On the other hand, there is a global trend toward the adoption of big data and data integration for decision-making (Zhang et al. 2011; Mukherjee and Kar 2017; Lim et al. 2023).

To enable the integration of data from disparate sources needed for precision horticulture, adoption of the FAIR principles by all parties is needed—regardless of

[1] https://www.agdatatransparent.com/principles/

the current state of data integrity—to enable data integration from disparate sources. In other words, we need to make sure that the collected data is findable, accessible, interoperable, and reusable, even if there is no immediate need or application for its integration. The investment we make today applying the FAIR principles will likely save considerable amounts of time and effort for future 'data-driven' precision horticulture.

This chapter explores six key elements or steps for successful and FAIR implementation of data-driven precision horticulture. The following sections consider planning, design, data collection, data management, data engineering, and data analysis as the key elements. We do not limit the discussion to examples or applications, although some sections benefit from specific case studies. Potential opportunities in the future and some concluding remarks are also provided.

10.2 The Six Elements of Data-Driven Precision Horticulture

Figure 10.1 shows the six key elements of data-driven precision horticulture with some example questions that should be answered at each step. We will review some conceptual and technical aspects in each element in the following sections.

Planning

Planning—the first element of data-driven precision horticulture—is about defining the purpose of collecting and analysing the data in question. This is important for building trust between the stakeholders. It is also about ensuring repeatability and reproducibility.

Our discussions with growers and stakeholder groups (pers. comm.) continually highlight the key importance of benefit-sharing in building trust between all parties. However, for this to come about, we must first address questions around whether data can, or should, be used or reused, before the data are even collected. Who gets the value? What measures are in place to ensure any benefits are shared fairly?

There is an increased expectation that communities or individuals should retain control over the data (Hummel et al. 2021), and potentially be able to opt-out of any process at any point. On the other hand, there is a lot of distrust of companies that use data without any benefit back to the contributors of that data (Chen 2020; Hummel et al. 2021).

To address these issues requires an understanding of who owns the data, who gives permission to use the data, what stakeholders have an interest in the use of the data, and what happens to the products of the use of the data. Thoroughly addressing and documenting the procedures for data sharing, including specifics on who will

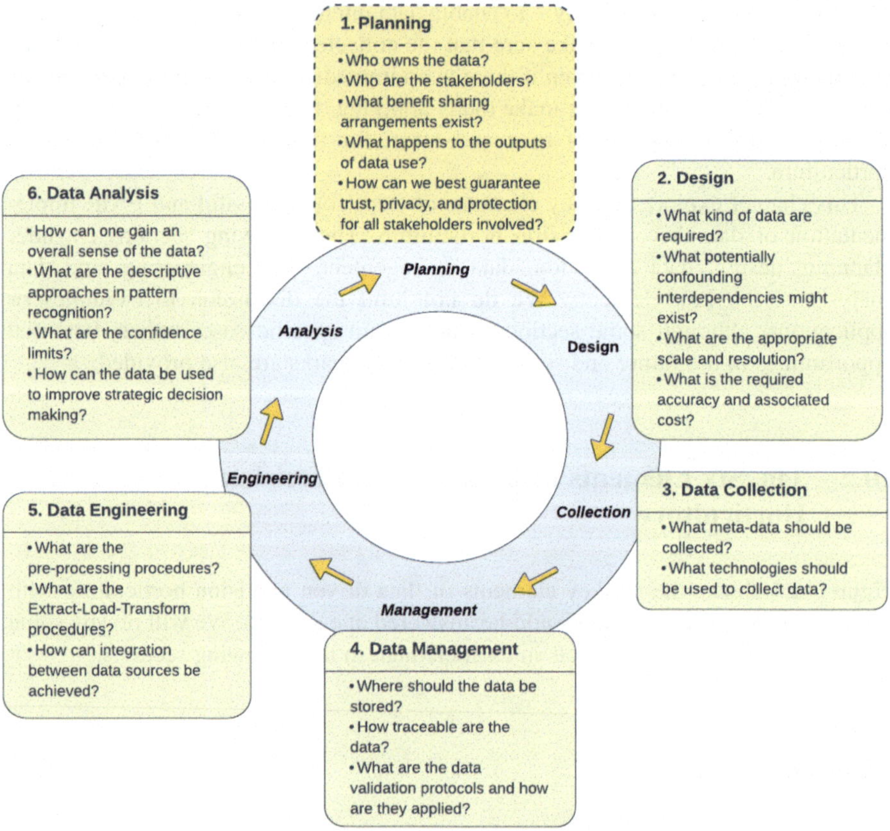

Fig. 10.1 The six elements of data-driven precision horticulture

access it, at what privacy level, and how it will be utilized, is crucial yet sensitive during the planning phase. These steps, while essential, can sometimes pose challenges. Stakeholders may have valid concerns regarding competitive advantages and legal compliance.

In addition to trust, there are two concepts that must be considered in planning data-driven precision. Firstly, the data collection process must be repeatable, i.e., it should be possible to collect the same data given the same conditions (allowing for a reasonable degree of variation beyond the experimental control e.g., as a result of natural variation within samples). Secondly, the processes involved in data cleaning and modelling must be reproducible; this means that given the same data, it should be possible to reproduce the findings at a different point in time. There are a lot of good guidelines for these practices (Sandve et al. 2013; Wilson et al. 2017).

The planning phase should also balance the investment in data collection and its precision against the cost of the data and minimise redundancy in collecting data that does not add information.

Finally, within the trust-repeatability-reproducibility framework, the context of the data must be known before the data are sharable. Understanding when, why, how and for what purpose the data were collected is essential metadata to be collected in its own right. This implies understanding the conditions under which the data can be used and re-used, and the key stakeholders involved. As mentioned previously, these are essential not just for good quality data products but the social license to enable the production of these data products.

Design

The design element of data-driven precision horticulture deals with practical aspects of data collection to achieve the planned outcome. The design of the project must be implemented according to parameters identified in the planning phase, to ensure trust, value protection, and cost-effectiveness.

Understanding what data to collect and the best way to collect it is not always a trivial task. The different elements of data-driven precision horticulture interact, which is why the elements shown in Fig. 10.1 are shown as an iterative cycle. There are many good statistical textbooks that deal with the design process e.g., (Whitlock and Schluter 2020). In the design phase, one should consider the issues set out in the following sub-sections.

Data Sources and Their Representations

Data sources can be extremely diverse and may require specific processing steps. Numerical sensors need calibration and validation, images need steps to extract information, genetic sequences often need considerable pre-processing. It may be that not every data type is needed for a particular question, but thought should be given if there is a possibility to bring them in later.

Sometimes what you want to know is not easy to measure. An example of this is fruit quality, which is a component of many variables. Fruit firmness or soluble solids may be some indication of quality but perhaps not capture the whole gambit. However, trying to capture all measures of quality is likely to be untenable and some proxies end up having to suffice.

Entities, Relationships, and Structures

Statistical modelling requires a key assumption of independence of the data that are being used to make predictions. Independence assumes that what is being observed is not influenced by the other observations, or at least the dependency is known and can be accounted for. For example, consider two plants in a row. They may respond in a similar manner not only due to the management but also due to underlying similarities (similar soil, water, microclimate characters). To disentangle questions of management we must fully understand the relationships between the objects or entities being measured. Examples include how an orchard is organised (rows, bays, plants), what are its biophysical characteristics such as as soils, slope, and presence or absence of shelter belts? Are there other factors that should be taken into consideration that may influence how plants grow? There are many useful tools to understand the structure of the data including entity-relationship data models and diagrams (Thalheim 2000), graph data models and diagrams (Bechberger and Perryman 2020; Holder 2021; Needham and Hodler 2019). These can be very useful when sharing data so the structures in the data can be understood.

Scale and Resolution

Another consideration is the multiple scales and resolutions of data that may occur in various dimensions. For example, data can vary across space and be collected at a national, regional, and block level. This is usually referred to as the spatial resolution. Temporal resolution regards to how often data are collected e.g., annually, monthly, daily, hourly. Within an object, the scale can vary from expression of a gene, phenotype of a leaf or leaves, measurements of whole plants or measurements across orchards. Collecting at too fine a scale can result in a data-wash that might not be necessary for the question at hand, but collecting at too coarse a scale may not provide sufficient resolution.

Accuracy, Precision, Mitigation and Validation

Measurement error is a key issue to consider throughout the process. All measurements come with a degree of error that ideally varies within a reasonable interval. This error can be random (noise) or non-random (biased). The design should be specific about measures to control the noise and increase precision (reduce bias), as well as strategies to take when noise variation is inevitably large. For example, will increasing the number of measurements enable a pattern to be discerned from the random errors around it and/or increase accuracy?

Finally, the design should involve thinking about ways in which issues and problems can creep into the data and mitigate them: plants can die, machines can fail, and batteries can run down. How will this affect the overall outcome?

Data Collection

Data collection is a critical element of data-driven precision horticulture that deals with the actual act of collecting data according to the plan and design. As data-driven horticultural precision is focussed on increasingly complex problems to tackle, technological load associated with data collection grows. This necessitates data moving from ad-hoc human collection to semi- or fully-automated collection. Technological solutions that allow easy data integration from disparate sources are the final step in the data maturity pipeline.

Planning for data collection should be done from several angles.

- Data collection and storage types: From ad-hoc manual to systematic manual, to semi-automated, to fully automated, to integrated.
- Quantification: From easily quantified e.g., length, to difficult to quantify e.g., disease symptoms.
- Metadata: From easily described e.g., length, to detailed descriptions e.g., disease symptoms.
- Lifetime: From defined lifetime to indefinite lifetime.

While the majority of horticulture sectors have historically collected data for one specific purpose, global trends in big data usage imply connection of historical data and provide data infrastructure that is designed to be used and reused for many different purposes. Broad data collection and integration is cumbersome due to the cost, in addition to the issues around trust mentioned in the section on planning. On the other hand, it enables post-hoc and virtual experiments and exploratory data analysis to find hidden relationships as well as more holistic modelling (see section on analysis). Data ontologies (Janssen et al. 2009) are extremely powerful tools that are becoming increasingly available for data integration and, if they are available for a crop, should be used to keep terminology consistent (Chuprikova et al. 2021).

Recent Technologies in Data Collection

Environment data such as incident radiation, soil-moisture, weather, and microclimate have a long history of collection and many off-the-shelf products are available for measuring these. Often a key consideration of these off-the-shelf products is the interoperability of the data: can they be easily combined and extracted? However, recently, new technologies have enabled further types of data collection. Technologies that measure plants, crops, and the environment that the plants and crops experience.

One of the most impactful recent technological advances across science has been the development of machine learning, particularly deep learning. As well as stimulating massive changes in data analysis, it has enabled collection of data from imagery using computer vision, enabling numerous opportunities for automating data collection. Deep learning computer vision can be employed with almost any source of imagery, including modern mobile phones, 3D cameras, videos, microscopes, hyper- and multi-spectral imagery, and remote sensing (aerial imagery and satellites).

Computer vision can make it easier to automate many common types of manual measurement, such as counting, sizing, or estimating colour profiles. It also enables new types of measurements, including biometric identification. Just as individual humans can now be identified by cell phones using facial recognition, biometric identification has now been trialled in sheep, in tree-ring patterns of logs, in spots and patterns on fish, and any other applications where individuals vary uniquely from each other. The technology that identifies and tracks individual fruit—either with a label or perhaps based on the patterns of pores, spots and stripes or hairs—has not yet been developed, but this is likely to be a key step in tracking fruit provenance along a value chain.

Many tasks for classifying an object in an image e.g., weed versus crop, counting e.g., fruit yield, or estimating area e.g., leaf area have recently moved from manual applications to automated processes using image or video, and there are likely to be many more as yet unthought of applications (Tian et al. 2020). A major area of research in computer vision work is occlusion—how to evaluate components that are hidden behind other structures e.g., fruit behind leaves (Chandel et al. 2015; Tian et al. 2020).

Hyper- and multi-spectral imaging are related technologies, which produce images of frequencies of light above, below and with much greater frequency resolution than those of standard cameras and the human eye. Many aspects of plant physiology are more apparent at specific frequencies, including chlorophyll content, nitrogen status, water content and biomass (Lu et al. 2020). Improvements in imagery size, cost, and spectral and spatial resolution, along with the required improvements in data storage and processing techniques, including machine learning, have enabled greater adoption of hyper- and multi-spectral imaging in horticulture. The reduction in sensor size has also enabled deployment by drone, which further increases the practicality of monitoring orchards.

Other emerging imaging technologies include 3D and thermal imaging. LiDAR, time-of-flight cameras and stereo cameras that capture three-dimensional spatial information as well as colour information, and allow for absolute spatial measurements such as length, area and volume and three-dimensional reconstructions of plants for physiological modelling. Thermal infrared cameras have been used for animal monitoring (Al-doski et al. 2016; Ishimwe et al. 2014) and plant disease detection (Yang et al. 2019). Active thermography adds a heat source, and allows sub-surface structures to be visualised, for example detecting the hollow made by a wood worm in a piece of wood (Al-doski et al. 2016).

At higher levels of digital maturity, data collection tends to happen in a more automated manner. Drones can be partially automated, following a programmed route, guided by high-accuracy global positioning systems. Permanent sensors placed in-orchard with solar panels and cellular connections can be fully autonomous but have limited sensing range. Sensors integrated with orchard machinery can collect data any time the machine is used in the orchard, but the data collection will be necessarily irregular and may have only partial coverage. Finally, robots may allow for full automation, but currently have limited capabilities and are prohibitively expensive. In each case, different sensor types are available. For example, contact sensors such as soil moisture can only be automated using a permanent sensor installation or robotics (López et al. 2011; van Tuijl et al. 2008).

Another type of sensing which can easily be automated is satellite imagery. Programmes such as NASA and United States Geological Survey's Landsat programme image the entire surface of the earth, with every point covered every 8 days. Satellites offer spatial resolutions up to 41 cm and a wide variety of spectral bands. Data can be accessed on-demand, and in some cases, free of charge. One limitation is that most satellite imagery cannot penetrate clouds. Synthetic Aperture Radar satellite imagery is not limited in this way, and has been used to measure canopy growth and crop type (Dingle Robertson et al. 2020).

Data Management

Data management is the element of data-driven precision horticulture concerned with the combination of regulations and actions for storing the collected data, ensuring findability, traceability, and interoperability according to the FAIR principles. The infrastructure and technologies used for data management must be carefully chosen according to the guidelines of the planning phase to ensure that the interests of data owners and stakeholders are protected.

As instruments and machines on-orchard or on-farm improve technologically, the amount of data collected (including the diversity of data as well as the total volume) is likely to grow almost exponentially. This leads to important considerations about data storage. There is a trade-off between how quickly data can be accessed for use and the cost of the storage of that data (Arutyunov 2012; Shamsi et al. 2013). There is a trend to move data away from on-site storage to specialised data 'cloud' solutions.

Trust is an important consideration for the location of data storage. Storage may also have data sovereignty or security issues, to be considered in the planning phase. Data are subject to the laws of the nation within which it is stored, rather than subject to the laws of the nation from which it was collected. In addition, data might also be subject to the laws of the nation under which the company operates e.g., the US CLOUD Act. These differences can potentially be in conflict and are part of a drive for cloud-computing companies to establish servers in more locations internationally (Chen 2020).

Moving data away from local storage to external storage providers also outsources the security of the system. This may be considered a positive or a negative depending on circumstances. Federated data governance systems are promoted as one way of increasing trust as they enable local storage of data with governance of the data and systems defined centrally (Latif et al. 2021). Security considerations are important, not only to ensure the privacy and integrity of the data, but also to prevent fraud. If data are shared across a supply-chain, trust in the data at each point is essential and methods to minimise or prevent fraud become increasingly essential.

Block-chain or distributed ledger solutions are often promoted as a way to ensure trust along a value chain (Bermeo-Almeida et al. 2018). Block-chain is a technology where information is stored on numerous servers across the internet (distributed ledgers) and so there is no single point that holds all information. Block-chains can be either permissionless, in that anyone can read the contents, or permissioned, in that permission is granted by the owner(s) of the blockchain. Because the information is distributed across the network, it becomes much more difficult to defraud (although permissioned block-chains may have fewer systems involved and so be easier to corrupt). However, because information is distributed across multiple nodes, it is not a solution for data-storage as the cost and time for distribution and verification of large quantities of data would be intractable. Rather, blockchain is championed to provide verification and digital tracing of pieces of information or data. While currently an immature technology for supply chains, there are ever-growing numbers of digital solutions utilising the technology. Some of the more mature examples for value chain tracking have been used to track pork (Xu et al. 2020), verify the provenance of wine (Danese et al. 2021), the organic status of wheat and oats (Xu et al. 2019) and security of infant formula (Rogerson and Parry 2020).

Data Engineering

Data engineering is the processing of data that involves any necessary pre-processing and filtering steps and then joining, merging and integrating disparate pieces of data. The goal of data engineering is to process and store all raw data collected in an earlier phase in a form such that it preserves the integrity of data and is efficient to be queried and sufficient to be used by the usage phase. The steps in data engineering are sometimes referred to as extract, transform and load.

As the data collection and automation becomes more advanced, data engineering can be a highly technical role that involves a series of steps. These can include the following.

- Loading (load raw data from storage).
- Cleaning (removing or fixing known errors).
- Filtering (removing irrelevant data for the purpose defined in the use of the data).
- Transform (convert units, change scales, split pieces of information etc.).
- Entity resolution (identifying data records that correspond to the same real-world objects). Sometimes this can be resolved by object identification (parsing data records and identifying objects by their unique IDs).
- Joins or linking (joining data for the same object coming from different data sources).

With data engineering comes a series of challenges to do these steps as computationally and time efficiently as possible and to overcome problems associated with use of different data platforms, different types of data, and different scales of information.

Data Analysis

Data analysis is the final element in data-driven precision horticulture and is intended to provide useful answers and insights. Primary interpretation of data concerns gaining a sense and understanding of it, which includes discovering patterns, clusters, or trends. Data visualisation is an effective tool for such *descriptive approaches* to data analysis.

The principles of effective data visualisation are one of the key fields in data science that was discussed in Telea (2014) and Midway (2020). Data visualisation techniques are usually fast in terms of computing time and range from simple figures to interactive desktop, mobile, or web applications, for historical, real-time or near-time data, such as interactive applications for genome-wide association studies (Hussain et al. 2018). While graphs and figures are usually methods to present raw data or summary statistics, analytical tools such as clustering or discriminant algorithms also provide insightful visuals that summarise highly dimensional data into clusters e.g., by examining differences in emissions of fruit after insect infestation (Mas et al. 2020).

Data understanding and description usually leads to a *predictive* approach to data analysis. We use the term 'predictive' in a broad sense that includes forecasting the future, estimation of the past or missing data, data smoothing and control. Predictive techniques are usually based on relating subsets of data, called variables, to one another by models. The variable(s) that is aimed to be predicted by the others is called the *response* or *output*. It is notable that the predictive model does not imply cause-effect relationship between the predictors and response; unobserved underlying physical principles, or a third confounding variable, may imply correlation between the predictors and response. For more information about differences between causation, correlation and confounding see (Stigler 2005).

Models are a fundamental part of prediction. Mathematically, a model is a parametric function that maps the response to the predictors. Figure 10.2 shows mathematical notation of models with meaning of each term, more information given in (Stigler 2005).

Empirical models, also known as phenomenological or non-mechanistic models, are mostly based on correlation between the response and predictors with few assumptions about the function and probability density of the residual. In this framework, some data sets of the predictors and response are collected

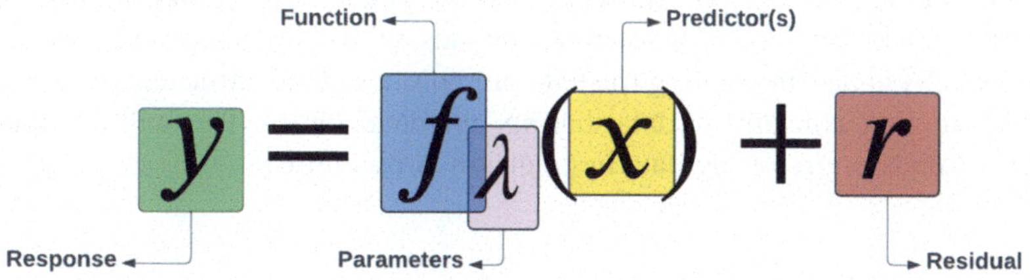

Fig. 10.2 Mathematical notation of models

simultaneously. The primary goal is estimation of the parameters and residual probability distribution based on the collected data sets. Neural networks (Kujawa and Niedbała 2021), machine learning (van Klompenburg et al. 2020) and deep learning (Ren et al. 2020) algorithms, either used in data collection or directly in the analysis phase, are all members of the empirical modelling family. These algorithms fit many model functions in several layers to find the best estimate. Natural Language Processing is another example that benefits from empirical modelling in text mining and social media analytics (Hamilton and Lahne 2020). The crucial point in empirical modelling is balancing the model to be sufficiently descriptive of the phenomena but not too specific and dependent on the collected data. For that purpose, model validation techniques are required (Parady et al. 2021).

Physical models, also known as mathematical or mechanistic models, are mostly based on the cause-effect relationship between the response and predictors. As a result, they usually make relatively more substantive assumptions about the function and parameters compared to the empirical methods and are designed specifically for one purpose e.g., biological switch models of fruit softening (van der Sman and Sanders 2012), or series of models employed in genetic regulatory systems (De Jong 2002). These models are generally more computationally complex and often involve a larger number of often unknown hyperparameters. However, if fully understood and validated, they can strongly predict the predictor-response systems with relatively smaller collected data sets and can be used for diagnostic purposes for qualitative changes in response behaviour.

It is worth noting that there is a spectrum from purely empirical to purely physical modelling.

Finally, a prescriptive approach to data usage focuses on decision-making and planning tools for actions that can change the outcome both qualitatively and quantitatively e.g., descriptive, and predictive approaches to data analysis show strong correlation between seasonal humidity and disease severity. As we move further in digital maturity, prescriptive approaches become more feasible.

10.3 The Future of Data-Driven Precision Horticulture

There are emerging opportunities in the data-driven precision horticulture space, enabled by continuing advances in data technologies as well as increasing awareness in the horticulture industry of the value that integrated data solutions can bring to informed decision-making. Meanwhile, the mission of organisations such as the World Data Systems[2] in creating trusted communities of scientific data repositories, are addressing the concerns of data sharing, and data control. This will facilitate the emerging trends in use of big data in precision horticulture.

[2] https://worlddatasystem.org/

Integrated data across the value chain provides many opportunities to create value for consumers. Much consumer data relating to the food system includes qualitative descriptions of tastes, smells and emotions elicited by the product. Historically the use of these data for predictive models has been limited by the qualitative nature. However, advances in Natural Language Processing enable a much more quantitative analysis. Linking genomic and chemistry data with descriptions of the consumer's experience provides an opportunity for understanding the requirements of different market segments. Linkage with genomic data enables accelerated breeding where key genomic areas are targeted or key crosses of parental material can be identified.

Advanced data integrated solutions provide a digital representation of a physical system or 'digital twin' (Mawson et al. 2022). Such digital twins have been used extensively in aircraft technology (Wang and Liu 2021), design of self-driving cars (Almeaibed et al. 2021), and in disaster and infrastructure planning for cities (Ruohomäki et al. 2018). Recently there has been a move for digital representations of horticultural and agricultural systems (Nasirahmadi and Hensel 2022; Sreedevi and Kumar 2020). While the predictions from a digital twin are only as good as the input data and models, they can be used to integrate data across the value chain and from genotype-to-phenotype to predict when, where and what crop should be planted as well as the outcomes of management decisions. Regardless of the type of models used in a digital twin, it should include a validation phase where predictions are compared continuously with actual outcomes on orchard or farm where feasible. This is made possible through internet of things (Friha et al. 2021) technologies where a digital twin can receive latest data from sensors. Otherwise, it has the deficiencies of many models in that predictions are only based on historical data and are not updated continuously.

Recent advances in generative AI and foundation models are quickly progressing many areas in data-driven precision horticulture. Chatbots are emerging for diverse use cases including data exploration and visualisation, and as coding and report writing tools. Foundation models are models trained on massive datasets for general tasks and can often be successfully applied to specific tasks using just a few examples of the desired output. The report from AI Infrastructure Alliance on Agents, Large Language Models, and Smart Apps (2023)[3] has a comprehensive review on these emerging technologies.

[3] https://ai-infrastructure.org/wp-content/uploads/2023/10/AIIA-AgentsLargeLanguageModelsandSmartApps-Oct3.pdf

10.4 Concluding Remarks

In this chapter, we considered the complexities that could be obstacles for data-driven precision horticulture. We argued that the initial cost and issues around data licensing often seem to be prohibitive despite the immense value that integrated data solutions might bring to the horticulture industry. However, there are three factors contributing to the emerging opportunities in this domain: Increasing awareness in the horticulture industry about the value of data-driven precision, the emerging international regulations and guidelines around data licensing, and the on-going advances in data science and technology, both in collection and analysis capacities.

Considering the global trends in the use of big data, we recommended that the six elements of data-driven precision aligned with the FAIR principles should be carefully considered in the design and planning for any data-driven horticultural project, regardless of the urgency or immediate vision for data sharing and reusage. We considered advancements in data collection, management, and analysis technologies, given some examples of recent employments of them in horticulture industry.

As technology advances, horticultural science and industry will find more opportunities and potential to create complex integrated solutions for their mid- and long-term planning. We argue that the advances in data technologies are changing our traditional understanding of data, in terms of its value and usage. In addition, as these advances are influencing our way to do science, they will subsequently affect our understanding of science. In other words, the road between horticultural science and data science runs both ways; as the demand in horticultural science makes use of advances in data science, the data science developments influence approaches to and perception of horticultural science and subsequently the industry.

Competing Interests The authors have no conflicts of interest to declare that are relevant to the content of this chapter.

References

Al-doski J, Mansor SB, Shafri HZBM, Zulhaidi H (2016) Thermal imaging for pests detecting - a review. Int J Agric For Plant 2:10–30

Almeaibed S, Al-Rubaye S, Tsourdos A, Avdelidis NP (2021) Digital twin analysis to promote safety and security in autonomous vehicles. IEEE Commun Stand Mag 5(1):40–46

Arutyunov VV (2012) Cloud computing: its history of development, modern state, and future considerations. Sci Tech Inf Process 39(3):173–178. https://doi.org/10.3103/S0147688212030082

Arya C (2021) Iot based precision farming and agriculture - aspects and technologies. Math Stat Eng Appl 70(2):1426–1433

Barbedo JGA (2020) Detecting and classifying pests in crops using proximal images and machine learning: a review. AI 1(2):312–328. https://doi.org/10.3390/ai1020021

Bechberger D, Perryman J (2020) Graph databases in action: examples in Gremlin. /z-wcorg/. http://public.eblib.com/choice/PublicFullRecord.aspx?p=6642795

Bermeo-Almeida O, Cardenas-Rodriguez M, Samaniego-Cobo T, Ferruzola-Gómez E, Cabezas-Cabezas R, Bazán-Vera W (2018) Blockchain in agriculture: a systematic literature review. In: Valencia-García R, Alcaraz-Mármol G, Del Cioppo-Morstadt J, Vera-Lucio N, Bucaram-Leverone M (eds) Technologies and innovation. Springer, pp 44–56. https://doi.org/10.1007/978-3-030-00940-3_4

Brini M (2020) Agri data regulation (EU GDPR and US). Enveve S.A. https://enveve.com/agri-data-regulation-eu-gdpr-and-us/

Chandel H, Shimla W, Vatta S, Shimla W (2015) Occlusion detection and handling: a review. Int J Comput Appl (0975–8887) 120(10):33–38

Chen A (2020) Shouting zeros and ones; digital technology, ethics and policy in New Zealand. Bridget Williams Books

Chuprikova E, Aguilar AM, Monsorno R (2021) An ontology-based visual analytics for apple variety testing. EGU21 Copernicus Meetings No. EGU21-15804. https://doi.org/10.5194/egusphere-egu21-15804

Codella J, Partovian C, Chang HY, Chen CH (2018) Data quality challenges for person-generated health and wellness data. IBM J Res Dev 62(1):3–1

Danese P, Mocellin R, Romano P (2021) Designing blockchain systems to prevent counterfeiting in wine supply chains: a multiple-case study. Int J Oper Prod Manag 41(13):1–33. https://doi.org/10.1108/IJOPM-12-2019-0781

De Jong H (2002) Modeling and simulation of genetic regulatory systems: a literature review. J Comput Biol 9(1):67–103

Dingle Robertson L, Davidson A, McNairn H, Hosseini M, Mitchell S, De Abelleyra D, Verón S, Cosh MH (2020) Synthetic aperture radar (SAR) image processing for operational space-based agriculture mapping. Int J Remote Sens 41(18):7112–7144. https://doi.org/10.1080/01431161.2020.1754494

Friha O, Ferrag MA, Shu L, Maglaras L, Wang X (2021) Internet of things for the future of smart agriculture: a comprehensive survey of emerging technologies. IEEE/CAA J Autom Sin *8*(4): 718–752

Hamilton LM, Lahne J (2020) Fast and automated sensory analysis: using natural language processing for descriptive lexicon development. Food Qual Prefer 83:103926. https://doi.org/10.1016/j.foodqual.2020.103926

Holder A (2021) Designing and building enterprise knowledge graphs. Morgan & Claypool. http://ieeexplore.ieee.org/document/9512460

Hummel P, Braun M, Tretter M, Dabrock P (2021) Data sovereignty: a review. Big Data Soc 8(1): 2053951720982012. https://doi.org/10.1177/2053951720982012

Hussain W, Campbell M, Walia H, Morota G (2018) ShinyAIM: shiny-based application of interactive Manhattan plots for longitudinal genome-wide association studies. Plant Direct 2: e00091

Ishimwe R, Abutaleb K, Ahmed F (2014) Applications of thermal imaging in agriculture — a review. Adv Remote Sens 03(03):128–140. https://doi.org/10.4236/ars.2014.33011

Janssen S, Ewert F, Li H, Athanasiadis IN, Wien JJF, Therond O et al (2009) Defining assessment projects and scenarios for policy support: use of ontology in integrated assessment and modelling. Environ Model Softw 24(12):1491–1500

Kujawa S, Niedbała G (2021) Artificial neural networks in agriculture. Agriculture 11(6):497. https://doi.org/10.3390/agriculture11060497

Latif R, Afzaal SH, Latif S (2021) A novel cloud management framework for trust establishment and evaluation in a federated cloud environment. J Supercomput 77(11):12537–12560. https://doi.org/10.1007/s11227-021-03775-8

Lim J-H, Kim J-H, Huh J-H (2023) Recent trends and proposed response strategies of international standards related to shipbuilding equipment big data integration platform. Qual Quant 57:863–884

López JA, Garcia-Sanchez A-J, Soto F, Iborra A, Garcia-Sanchez F, Garcia-Haro J (2011) Design and validation of a wireless sensor network architecture for precision horticulture applications. Precis Agric 12(2):280–295. https://doi.org/10.1007/s11119-010-9178-1

Lu B, Dao PD, Liu J, He Y, Shang J (2020) Recent advances of hyperspectral imaging technology and applications in agriculture. Remote Sens 12(16):2659. https://doi.org/10.3390/rs12162659

Mas F, Horner R, Cazeres S, Alavi M, Suckling DM (2020) Odorant-based detection and discrimination of two economic pests in export apples. J Econ Entomol 113:134–143

Mawson AJ, Stanley CJ, Zhu J, Pattemore DE, Chooi KM, Oliver RJ, Lin HT, Harker FR (2022) Developing a digital twin of apple production and supply chain ecosystems. In: XXXI international horticultural congress (IHC2022): III international symposium on mechanization, Precision Horticulture 1360, pp 129–136

Midway SR (2020) Principles of effective data visualization. Patterns 1(9):100141

Mukherjee R, Kar P (2017) A comparative review of data warehousing ETL tools with new trends and industry insight. In: IEEE 7th international advance Computing conference (IACC). IEEE, pp 943–948

Nasirahmadi A, Hensel O (2022) Toward the next generation of digitalization in agriculture based on digital twin paradigm. *Sensors 22*(2):498

Needham M, Hodler A (2019) Graph algorithms

Olayinka AS, Adetunji CO, Nwanjwo W, Olugbemi OT, Olayinka TC (2022) A study on the application of Bayesian learning and decision trees IoT-enabled system in postharvest storage. In: Artificial intelligence-based internet of things systems. Springer, Cham, pp 467–491

Parady G, Ory D, Walker J (2021) The overreliance on statistical goodness-of-fit and under-reliance on model validation in discrete choice models: a review of validation practices in the transportation academic literature. J Choice Model 38:100257

Plazas JE, Bimonte S, De Sousa G, Corrales JC (2019) Data-centric UML profile for wireless sensors: application to smart farming. Int J Agric Environ Inf Syst 10(2):21–48

Ren C, Kim D-K, Jeong D (2020) A survey of deep learning in agriculture: techniques and their applications. J Inf Process Syst 16(5):1015–1033. https://doi.org/10.3745/JIPS.04.0187

Rogerson M, Parry GC (2020) Blockchain: case studies in food supply chain visibility. Supply Chain Manag Int J 25(5):601–614. https://doi.org/10.1108/SCM-08-2019-0300

Ruohomäki T, Airaksinen E, Huuska P, Kesäniemi O, Martikka M, Suomisto J (2018) Smart city platform enabling digital twin. In: 2018 international conference on intelligent systems (IS). IEEE, pp 155–161

Sandve GK, Nekrutenko A, Taylor J, Hovig E (2013) Ten simple rules for reproducible computational research. PLoS Comput Biol 9(10):e1003285. https://doi.org/10.1371/journal.pcbi.1003285

Sarker MN, Islam MS, Ali MA, Islam MS, Salam MA, Mahmud SH (2019) Promoting digital agriculture through big data for sustainable farm management. Int J Innov Appl Stud 25(4):1235–1240

Shamsi J, Khojaye MA, Qasmi MA (2013) Data-intensive cloud computing: requirements, expectations, challenges, and solutions. J Grid Comput 11(2):281–310. https://doi.org/10.1007/s10723-013-9255-6

Sreedevi TR, Kumar MS (2020) Digital twin in smart farming: a categorical literature review and exploring possibilities in hydroponics. In: 2020 Advanced Computing and Communication Technologies for High Performance Applications (ACCTHPA), pp 120–124

Stigler SM (2005) Correlation and causation: a comment. Perspect Biol Med 48(1 Suppl):S88–S94

Telea AC (2014) Data visualization: principles and practice. CRC Press

Thalheim B (2000) Entity-Relationship Modeling: Foundations of Database Technology. Springer, Berlin/Heidelberg

Tian H, Wang T, Liu Y, Qiao X, Li Y (2020) Computer vision technology in agricultural automation — a review. Inf Process Agric 7(1):1–19. https://doi.org/10.1016/j.inpa.2019.09.006

van der Sman RGM, Sanders M (2012) Prediction of postharvest firmness of apple using biological switch model. J Theor Biol 310:239–248 van Klompenburg T, Kassahun A, Catal C (2020) Crop yield prediction using machine learning: a systematic literature review. Comput Electron Agric 177:105709. https://doi.org/10.1016/j.compag.2020.105709

van Tuijl B, van Os E, van Henten E (2008) Wireless Sensor Networks: State of the art and future perspective. Acta Hortic 801:547–554. https://doi.org/10.17660/ActaHortic.2008.801.61

Wang T, Liu Z (2021) Digital twin and its application for the maintenance of aircraft. In: Handbook of nondestructive evaluation 4.0, pp 1–19

Whitlock MC, Schluter D (2020) The analysis of biological data. Macmillan Learning. https://books.google.co.nz/books?id=i0DMywEACAAJ

Wilkinson MD, Dumontier M, Aalbersberg IJ, Appleton G, Axton M, Baak A, Blomberg N, Boiten J-W, da Silva Santos LB, Bourne PE, Bouwman J, Brookes AJ, Clark T, Crosa M et al (2016) The FAIR guiding principles for scientific data management and stewardship. Sci Data 3: 160018

Wilson G, Bryan J, Cranston K, Kitzes J, Nederbragt L, Teal TK (2017) Good enough practices in scientific computing. PLoS Comput Biol 13(6):e1005510. https://doi.org/10.1371/journal.pcbi.1005510

Wiseman L, Sanderson J, Zhang A, Jakku E (2019) Farmers and their data: an examination of farmers' reluctance to share their data through the lens of the laws impacting smart farming. NJAS Wageningen J Life Sci 90–91:100301. https://doi.org/10.1016/j.njas.2019.04.007

Xu X, Weber I, Staples M (2019) Case study: AgriDigital. In: Xu X, Weber I, Staples M (eds) Architecture for Blockchain applications. Springer, pp 239–255. https://doi.org/10.1007/978-3-030-03035-3_12

Xu J, Guo S, Xie D, Yan Y (2020) Blockchain: a new safeguard for agri-foods. Artif Intell Agric 4: 153–161. https://doi.org/10.1016/j.aiia.2020.08.002

Yang N, Yuan M, Wang P, Zhang R, Sun J, Mao H (2019) Tea diseases detection based on fast infrared thermal image processing technology. J Sci Food Agric 99(7):3459–3466. https://doi.org/10.1002/jsfa.9564

Zhang Z, Bajic VB, Yu J, Cheung K-H, Townsend JP (2011) Data integration in bioinformatics: current efforts and challenges. In: Bioinformatics trends and methodologies. InTech, pp 41–56

Index

P. Newell Price et al. (eds.), *Precision Agriculture in Field Horticulture*, Progress in Precision Agriculture, https://doi.org/10.1007/978-3-031-96534-0

The manufacturer's authorised representative in the EU is Springer
Nature Customer Service Centre GmbH, Europaplatz 3, 69115 Heidelberg,
Germany. If you have any concerns regarding our products, please
contact ProductSafety@springernature.com

Printed and bound by CPI Group (UK) Ltd, Croydon, CR0 4YY
30/09/2026
02225590-0008